BIRDS THROUGH INDIGENOUS EYES

BIRDS THROUGH INDIGENOUS EYES

*Native Perspectives on
Birds of the Eastern Woodlands*

DENNIS GAFFIN

with the collaboration of

MICHAEL BASTINE
& JOHN VOLPE

PRINCETON UNIVERSITY PRESS
PRINCETON & OXFORD

Published by Princeton University Press
41 William Street, Princeton, New Jersey 08540
99 Banbury Road, Oxford OX2 6JX

press.princeton.edu

All Rights Reserved

Library of Congress Cataloging-in-Publication Data

Names: Gaffin, Dennis, author. | Bastine, Michael, interviewee. | Volpe, John
    (Wildlife rehabilitator), interviewee.
Title: Birds through indigenous eyes : Native perspectives on birds of the eastern woodlands /
    Dennis Gaffin ; with the collaboration of Michael Bastine and John Volpe.
Description: Princeton : Princeton University Press, [2024] | Includes bibliographical references.
Identifiers: LCCN 2023030653 (print) | LCCN 2023030654 (ebook) | ISBN 9780691250847
    (hardback) | ISBN 9780691250908 (ebook)
Subjects: LCSH: Birds—East (U.S.) | Algonquin Indians—Ethnozoology. | Ojibwa Indians—
    Ethnozoology. | Traditional ecological knowledge—East (U.S.) | BISAC: NATURE / Animals /
    Birds | SCIENCE / Life Sciences / Zoology / Ornithology | LCGFT: Interviews.
Classification: LCC E99.A35 G34 2024 (print) | LCC E99.A35 (ebook) |
    DDC 598.0974—dc23/eng/20231024
LC record available at https://lccn.loc.gov/2023030653
LC ebook record available at https://lccn.loc.gov/2023030654

British Library Cataloging-in-Publication Data is available

Editorial: Fred Appel and James Collier
Production Editorial: Sara Lerner
Text and Jacket Design: Chris Ferrante
Production: Erin Suydam
Publicity: Matthew Taylor and Carmen Jimenez
Copyeditor: Jennifer Harris

Jacket Credit: Norval Morrisseau, *Serenity*. Permissions granted and authenticity ensured by the Norval Morrisseau Estate Ltd, OfficialMorrisseau.com. Anishinaabe artist Norval Morrisseau (1931–2007) is considered by many to be the grandfather of contemporary Indigenous art and is famous for shattering societal, racial, and sexual prejudices and stereotypes in the 1960s. Morrisseau, creator of the Woodland School of art, is best known for using bright colors and thick dark lines and for portraying traditional stories, spiritual themes, and political messages.

This book has been composed in Karol and Karol Sans

Printed in the United States of America

10 9 8 7 6 5 4 3 2 1

CONTENTS

List of Illustrations	vii
Acknowledgments	ix
Introduction	I
John Volpe	4
Michael Bastine	7
Chapter One. View from Above	15
The Context: Physical and Metaphysical	24
Chapter Two. Bird Messengers, Totems, Lessons	51
General Philosophy and Practice	78
Chapter Three. Messages of Color	91
Color and Spirit	93
Chapter Four. Natural Law and Original Instructions: Consequences, Changes, Connections	119
Conclusion. Awareness and Natural Law	137
Before the Tobacco Offering	140
Afterthoughts	145
Notes	153
Suggested Reading	157

The bird illustrations are available in color via links at
https://press.princeton.edu/isbn/9780691250847

Ceremonial Pipes	13
Eastern Phoebe	23
Northern Flicker	39
Belted Kingfisher	55
Gray Catbird	59
Cedar Waxwing	97
House Sparrow	109
Black-Capped Chickadee	115

ACKNOWLEDGMENTS

Gratitude to the State University of New York for partial funding of this project and to Rebekah Leith for many interview transcriptions.

I especially thank colleagues and friends Michael Bastine and John Volpe for many, many hours of discussion, and Ana Bodnar for her loving support.

Many thanks also go to the staff at Princeton University Press, especially Fred Appel for his gracious nurturing of this book.

BIRDS THROUGH INDIGENOUS EYES

My two Indigenous conversation partners are Michael (Mike) Bastine, of Algonquin (Algonkian) descent, a member of the Kitigan Zibi Anishinabeg First Nation of western Quebec, of the Bear Clan, and a resident of Erie County, New York; and John Volpe, of Ojibwe and Métis descent, a member of the Nipissing First Nation of Ontario, of the Turtle Clan, and resident of Wyoming County, New York. Mike is a former apprentice of the Tuscarora medicine man Wallace "Mad Bear" Anderson, of the Tonawanda reservation of western New York, described in Boyd's (1994) *Mad Bear: Spirit, Healing and the Sacred in the Life of a Native American Medicine Man*, in which Mike appears. (The Algonkian nations have lived in territories now part of Quebec, Ontario, and northern New York. The Ojibwe, or Ojibwa or Ojibway, have historically populated what is now the north central United States, such as Wisconsin and Minnesota, and north into what is now southern Ontario and Manitoba, and farther west.)

This book, the fruit of many, many hours of discussion, conveys the understanding and experience of Mike and John with Eastern Woodlands birds and nature. Their upbringings, teachings from Native elders, significant immersion in nature, and extensive bird and animal rehabilitation experience comprise the backbone of the verbatim interviews throughout the book.

This book is intended neither as a guide to all Native American or First Nations accounts of or experiences with each bird species of the Eastern Woodlands, nor as a comprehensive inventory of all Indigenous stories and understandings of such birds. While some traditional Native stories about birds are recounted here, the aim is more to present general perspectives and worldviews as they manifest in the words and experiences of Mike and John. In this way, the information and perspectives appear as living things, rather than as static intellectual artifacts to be examined, catalogued, and shelved.

Interspersed between my verbatim discussions with Mike and John, typically conducted around John's kitchen table, I have inserted excerpts from my own observations of birds and analysis of my bird encounters, in italics. In and through the verbatim dialogue and journal extracts, the same set of North American Eastern Woodland bird species is continually referenced. This book features illustrations and descriptions of seven of these: the eastern phoebe, northern flicker, belted kingfisher, gray catbird, cedar waxwing, house sparrow, and black-capped chickadee.

The reader will note that my italicized journal extracts reflect a view of birds that is both deeply personal and rooted in a mostly Western understanding of birds. Those sharing this cultural stance typically approach birds as beings who feed one's aesthetic sensibilities. One takes pleasure, and sometimes displeasure, in their color, shape, movement, call, song, and behavior. How could one not? After all, birds are natural practitioners of song, opera, ballet, and drama, especially in territorial and mating behavior.

I have some bird memories and imaginings of my birding past that stand out as important markers of local life, a nourishing repertoire of key life events. I easily can conjure up in my mind from twenty years ago the indigo bunting—a purple-blue finch—when he flew along the treed side of the old railroad bed by the seven-acre pasture. His unparalleled color and once-or-twice-in-a-lifetime sighting enchanted. Then there was the time the red-tailed hawk, for his own fowl dinner, clawed and snatched away the blue jay. Of course, the catbird whines and whines by the side door. . . . And, when traveling in foreign lands, I distinctly remember, among others, the Arctic skuas and the puffins on the Faroe Islands, the honeycreeper in Hawai'i, the bald eagles at Haida Gwaii, the pink flamingos in Cuba, the laughing falcon in Costa Rica, the frigate bird in Guadeloupe, all of whom are still etched in my mind.

Everyone, to one degree or another, along some continuum of awareness and experience, has a personal ornithology, which invariably draws on the perspectives and classifications shared within one's culture—a "folk" understanding of birds one shares with others in close cultural proximity. Anthropologists and other scholars who investigate these shared understandings practice *ethno-ornithology*. This includes but is not limited to the study of the ways a group "uses" birds, literally or symbolically, and how individuals look at birds, in every sense of the word "look," including which features of a bird are most salient or important.

Individuals can modify their personal and cultural understandings of birds in their lifetimes. So although Mike and John have been deeply shaped by their Native heritages, their views are not simply representations of the views of birds, and of nature, of their respective Native communities. As for me, my in-depth exposure to the personal and cultural orientations of Mike and John—through our extended discussions and shared walking-in-nature experiences—has deeply influenced my own views. (For further discussion of the author's birdwatching journey see the afterthoughts.)

It is my personal and scholarly belief that reproducing verbatim discussion, as first-hand data, can serve as a corrective to our tendency to overanalyze and can help us more accurately connect with the mindset and experience of everyday people—who merit recognition as experts in their own right. Moreover, reproducing the voices of Native interlocutors is more in keeping with traditional Native patterns of communication and teaching in spoken word and stories.

JOHN VOLPE

As noted earlier, almost all the interviews and discussions in the following chapters took place at a kitchen table in John's house, at his homestead and animal sanctuary. John typically has many live-in animal companions, including owls, seagulls, songbirds, injured raccoons or other mammals, and many turtles. Throughout the house, one also finds bird carvings and other artwork, and items of natural history. Often on the kitchen island counter is a cage with a bird recovering from an injury or a fish tank hosting injured or young turtles. On one occasion, early on in our talks, on the counter were three large plastic bowls with water, each with a turtle of a different species: the painted, the wood, and the snapping turtles.

John grew up spending most of his waking time outdoors. His great grandfather was Algonquin—as John says, "He was supposed to have been a Chief, way back then." Although John had Native teachings from his mother, Seneca Chief Skye, and other teachers, he says that his knowledge "overall has come from being out in the woods." He is a sweat lodge leader, a grower and user of cere- monial tobacco, a woodcarver, an artist, and a former professional taxidermist. He is an animal rehabilitator and an environmental monitor and activist, a nature educator to local schools, and a consultant to New York State's Department of Environmental Conservation (DEC).

For many years, locals have brought him injured or dead birds, deer, or other mammals. John is learned in ecology and animal biology and is self-taught in veterinary medicine, especially in the treatment of wounded birds. From a single feather, he can tell a bird species, what kind of feather it is (from which part of the bird), and whether it is from the right or left side. He also is learned in the physical and metaphysical connections among and between animals and humans. He is knowledgeable in the ways in which individual animals or specific species (alive or recently deceased) align with specific humans and how birds and animals can provide clues to an individual's own issues, concerns, and problems. Depending upon the bird, how s/he was injured and/ or how s/he died, and who brought the bird to him, he can read issues and problems of the person involved.

Bird, deer, and other animal traits and spirits inform John and sometimes impel him to provide counsel and advice to the person who brings him the animal. He can read deer antler shape, size, and positioning as diagnostic of medical and psychological issues of the human who killed or found him. For example, he might advise a man to develop his feminine side for more emotional balance, with a view to helping him improve his personal relationship with his life companion.

While familiar with Native traditions of omens and totems, and the sacred in Nature, John has also developed his own perspectives and practices as a consequence of deep immersion and first-hand experience. Although he learned from Seneca Chief Skye about reading signs from parts of birds, right and left, no one taught him about reading deer antlers. He learned it during many years of working with animals, alive and dead: as he says, he "just picked it up from Nature."

JOHN: I spent all my time in the woods when I was young and growing up. All my time. That was the only thing I could really relate to.

DENNIS: As a child did you learn stories about birds, their calls and songs, their meaning?

JOHN: I didn't learn those things till way later, not until I really started walking the path. That's when I started to look at the stories about birds and the stories the birds tell. I wanted my life to have those types of values that they teach. That's what I love about the stories: they always have deep values.

Here John describes some of his own background:

JOHN: My mother used to talk about Nipissing [Ontario], where her mother came from. She would talk about how the women went up to the water's edge, when the men went hunting or fishing. . . . They used to go up with stuff in their hands, tobacco, and make offerings to the guys as they were going out in their canoes. And then she talked about picking berries with everybody, where they would go up with their baskets, and pick berries. . . . My mother was forced into a residential school.

MICHAEL BASTINE

Mike is a frequently sought out speaker and healer. He has been a bird and animal rehabilitator, notably training volunteers how to work with wounded and captured birds, especially raptors such as hawks and eagles. He also raises and utilizes ceremonial tobacco. Mike has been a long-time traditional teacher and leader in various locales, traveling in the East and Midwest of the United States to join Native gatherings. Co-author of *Iroquois Supernatural*,[1] he has long been a frequent guest speaker at the State University of New York College at Buffalo.

MIKE: You know, I didn't know enough in the early part of my life. Sometimes now, when I offer tobacco, recollecting some things when I was young, I say I'm not going to do that unconnected-with-Nature behavior again. That's in part why I feed the chickadees and why I take care of the birds. It makes a lot of sense that as we go through life we should be learning and finding more and more connections.

It is not only the Native culture that has informed me, but I guess you could say that Nature was my playground. It was my school. It was where I found comfort in life, playing outside with frogs, with Nature.

I asked Mike about his teachers and heritage:

MIKE: Mad Bear was my main teacher. He was a powerful Tuscarora medicine man of the Bear Clan who lived on the Tonawanda reservation. The people on the reservation went to him for healing and advice. And even some white people came onto the reservation to see him. That is somewhat written about in Boyd's book *Mad Bear*.[2] For decades plus, I have been carrying on his teachings.

DENNIS: You also said Ted Williams was another major influence, major teacher, in your life.

MIKE: Yes. He was Tuscarora, also. Wolf Clan. He was the son of a chief and medicine man. He wrote a book, *The Reservation*,[3] about life on the reservation. Later in life he moved to North Carolina, where the Tuscarora people had lived for many generations.

DENNIS: So you would say that Mad Bear and Ted Williams were your major teachers?

MIKE: Yes, they were the main ones. Through them you meet other teachers on the journey. But they were the two main influences. You kind of need somebody like that, more than just dressing like [Native], changing your hair style, and saying things.

Those two were really big influences. I guess it is a form of confirmation, always exchanging with them. Not that you need to have their approval, but if you did not have their approval, you would no longer be in their company. And you learn from them. And if you want to keep hanging out with these guys, you have to be up on the same kind of concept perception that says, "Wow, that makes a lot of sense now."

Because Ted would do things that would be unorthodox. Ted was even more unorthodox among the traditional. But they liked that because he was perceived as, acknowledged as, a Heyoka. It's a high ranking, and it is referred to as Coyote or Clown. It is integrating humor with sacred things. So, without being disrespectful, he would have fun within the sacredness of the gathering of elders.

Ted's dad was a Chief. But Ted never accepted chieftainship. But he was a Sacred Clown. Ted was really good at that. At lightening things up.

DENNIS: Tell me about the golden eagle wing.

MIKE: I got a golden eagle wing, and also a very special feather. It is a totem feather, and it has faces right in the feather. It has profiles of a face.

The wing came from Spotted Eagle, Rolling Thunder's son. I have been with Rolling Thunder; he knew who I was. . . . We were in Ohio. RT was doing a presentation at Kent State when I was young and learning from Mad Bear. My mom and dad and I drove down, my brother, my sister. We all went down to see Rolling Thunder. . . . Rolling Thunder pulled me aside, asked me, "Would you like to come out to Metatate?" a community that he was starting, with traditional structure and teachings. But I said, "Who is going to help Mad Bear?" Then he stopped because he knew. . . .

You don't make an agreement in writing. I didn't say to Mad Bear that I would be his servant for as long as he was here. You just do it. You don't ask if help is needed. You just help. That was acknowledged, and Rolling Thunder saw that. He knew.

Because at the time, in your early twenties, you think you know. But you really do not know much. You learn, not just by listening, and asking questions. But you also just observe. There were great things and Mad Bear would address that. And I would go, "Oh, man." I never saw him get stumped in any situation. He always had a way to help people deal with whatever the circumstances were.

Mad Bear always operated from a spiritual perspective, looking at Nature as a sacred doctrine, a sacred Teaching. You pick up from that. You say, "Wow, he's right." And he would do offerings before they would cut a tree or disturb some of the land. You knew that he understood there's more than just getting a shovel or moving things around. There are living things that you are going to disturb. So . . . In fact, I remember one time, we were out in Arizona. We had driven out there. And he was going up to the car, to the windshield, with sage. And he had tobacco. And I said, "What are you doing?" He said, "See all these bugs on the windshield, they all gave their lives so we could get here." And I am going to myself, "Who thinks like that?" You never hear about anybody acknowledging bugs on a windshield, that they gave their lives so we could get there. . . . There's this understanding that these are living things, and that yes, there is this vehicle we needed to get there.

We try not to, but how do you steer around bugs?
[John laughs.]

A few more words are in order about Mike and John before we begin.

They are both advocates and activists on local environmental issues. But to call them environmentalists—a Western term— implies that the "environment" or even "ecology" is separate from other aspects of life, including spirituality. They know that the physical environs and nonhuman living things are not an aspect of life. They are life. It is more that they are advocates for Nature and Natural Law, and for the Spirit in Nature.

Both John and Mike's Native practices were partly dormant before adulthood, in great part a result of how the dominant Euro-American, white culture had tried to destroy much of their parents' and ancestors' Native languages and heritage. Their somewhat similar personal and cultural histories of reactivating their most Indigenous selves demonstrate how, as one ages, one can evolve and learn to find greater meaning in one's heritage, and creatively add to it.

Neither John nor Mike hold official positions in a tribal or band council. Yet they are called upon to perform Native ceremonies in New York state and beyond. John regularly hosts a sweat lodge on his property and has conducted traditional ceremonies (which often include species-specific bird feathers), such as when he led a Seneca ceremony for the repatriation of Native bones from central New York state.

People might consider both Mike and John "elders," in the traditional Native sense as "wisdom-keepers." Yet Mike and John continue to follow the lead of Mike's great mentor, Mad Bear, where the use of titles or honorifics is concerned. Natives often sought Mad Bear for healing, medical, and spiritual assistance, and he was widely regarded as a medicine man. But he never referred to himself as a medicine man or elder. As Mike explains, "we should not accept a title until we are in our seventies or eighties. Humility is to be practiced at every stage of life." Thus, neither Mike nor John call themselves elders. (Mike says, "The animals regard John as an elder.") As a consequence, I do not call them elders either in this book. Although they do not speak on behalf of a specific tribal community, both are certainly culturally situated carriers and practitioners of Haudenosaunee (Iroquois), Algonkian, and Ojibwe traditions and worldviews.

After John read the first draft of this book, he requested that where I invoked the word "religion" in association with his or any Native American perspective, I use, instead, words such as spirit or nature. I agreed, even though many readers may find some of the views of Mike and John "religious." Modern Western systems of thought tend to conceive of reality as bifurcated between things tangible and intangible, physics and metaphysics, rational and irrational, natural and supernatural, and science and religion. I never heard John or Mike use the words mystical, metaphysical, irrational, or supernatural.

During the global COVID pandemic that began in March 2020, when hundreds of millions of people were confined to their homes, there appeared many newspaper and magazine articles on the importance of birds in people's lives. During and because of the pandemic, birds became more appreciated as beings crucial to our sanity. One *New York Times Magazine* writer proclaims how birds came to be meaningful markers, arbiters, of time itself:[4]

> Everything is on hold. The future is indeterminate. We do not know what will happen next. We cannot. The sparrows that hop on the bricks of my backyard wall have daily routines I am coming to know, and witnessing them is calming to watch when I have few of my own. As I look out of my window in lockdown, my attention is fixed on these birds, rather than trees or distant rooftops, because I am desperate for novelty, to watch things that alter; for in seeing change, I can parse time.

Citing her reading of the writings of birdwatchers in prison, she found that

> These men were writing trajectories, borrowed from the lives of birds, that made the passage of time meaningful. . . . I began to see that what these men did was a form of devotion. They were using the small lives of birds as things they could orient themselves against. Their patient observations remind me of how monks in the medieval monasteries ordered their days to fill them with meaning. . . . From one place we can witness the sweep and dip of the universe about us.

My own long-standing immersion among birds in my home in the countryside and my association with Mike and John inspired me to write this book—to provide readers who may be unfamiliar with Indigenous worldviews a grounded sense of how this growing realization of the importance of birds has been present in Indigenous knowledge for eons. (For further discussion of ethno-ornithology, beliefs, and mysticism, see the afterthoughts.)

CEREMONIAL PIPES

A broken bird-head pipe (*top*) and two pipe bowls (*middle & bottom*)

The pipes depicted here, now museum artifacts, are self-directed ceremonial sacred tobacco pipes of Algonquin, Haudenosaunee (Iroquois), or Ojibwe origin. The sacred pipes are self-directed—meaning that the "head" of the pipe points to the smoker, not to the observer as is usual in Western and other pipes. This indicates, as has been described in W. Engelbrecht's book *Iroquoia*,* that in sacred pipe smoking the smoker could commune with (the image of) the bird in spiritual interchange, especially as tobacco or other products can produce altered states of consciousness.

Credits: (*top*) Pipe bowl. Canadian Museum of History. VIII-F:8502, IMG2013-0158-0017. (*middle & bottom*) Various pipe bowls. Canadian History Hall, Canadian Museum of History, IMG2017-0089-0004. Middle: VIII-F:8496, VIII-F:23308 a-b, VIII-F:8504. Bottom: VIII-F:8434, VIII-F:8502, VIII-F:23309.

* Engelbrecht (2003: 55–56).

CHAPTER ONE

VIEW FROM ABOVE

ME: Why am I alive?

OLD WOMAN: Because everything else is.

ME: No, I mean the purpose.

OLD WOMAN: That is the purpose. To learn about your relatives.

ME: My family?

OLD WOMAN: Yes. The moon, stars, rocks, trees, plants, water, insects, birds, mammals. Your whole family. Learn about that relationship. How you're moving through time and space together. That's why you're alive.

—RICHARD WAGAMESE, *Embers: One Ojibway's Meditations*, 2017

We are all familiar with modern scientific approaches to why birds matter, which generally stress the roles that birds play for sustainable environments—birds disperse seeds, pollinate plants, and reduce populations of insects injurious to plant growth and so on—and which also emphasize the theme of how immersion in nature and with birds is good for human mental health. (Nature-deficit disorder is increasingly becoming recognized in human psychology.[1]) Another very common way birds appear in the Western imagination is as a symbol—often a symbol of a common human personality trait considered either praiseworthy or deserving of criticism. Certain bird species are thought of as brave (blue jay), pretty (yellow warbler), talkative (chickadee), vulgar (vulture), dirty (seagull), melodious (mockingbird), irksome (gray catbird), or cowardly (chicken). All of these are associated with the bird species as a whole. The species in question is understood, reductively, in terms of that one trait. The salient feature supposedly incarnated by the type of bird in question reflects widely shared cultural priorities and preoccupations.

This reductionist approach can be seen in the symbolic meaning that the founders of the United States bestowed on the bald eagle. Like other birds of prey, this species has excellent vision and is a superb predator and hunter of fish and small animals. The decision of the founding fathers to elevate this bird's predator skills parallels the white colonists' predation of Indigenous people, animals, and the environment.

Mike and John perceive this species in a different light.

MIKE: Why do Natives look to eagles and hawks and hold them in such high esteem? It is not because they have great vision or sharp talons or a beak that can tear flesh. It is because of the eagles' interactions with crows and red-tailed hawks—how the eagles and other birds of prey restrain themselves from hurting crows and others that are bothering them. Often in the sky you can see smaller birds, like crows or starlings, attacking eagles and hawks. The eagles and hawks do not react and do not retaliate with the crow, or starlings, or other birds that bother them. If they so desired eagles could easily, even in midair, use their strong talons to injure or kill the birds chasing and bothering them. Thus, the message that should be sent out is—no retaliation or violence when annoyed, when one's life is not in jeopardy. Restraint should be used with weapons—talons or guns. Weapons should only be used to feed yourself.

We should think highly of eagles, not for their size or strength, but because they restrain themselves. They do not go around with weapons, badges, or guns thinking to keep law and order or hassling weaker species.

And I bring out to people: Do you think that the hawk hates the rabbit and that is why he is killing it? Or he hates the squirrel? He loves them, and his own family, and that's why he is taking it, because it is going to keep his family alive. There's no emotional state here. This is about survival. They know that this is how life should be. It is the humans who keep skewing the perceptions and twisting them. Whether it is for religious or political ends, it doesn't matter much.

DENNIS: I have a question. The kingbird, you know, the eastern kingbird. I would say, from a human perspective, is not nice.

MIKE: Well, from a human perspective.

DENNIS: From a human perspective because he tends to attack other birds. Is that what your experience is with the kingbird?

JOHN: Yes, I know the kingbird. . . . He's no different than a blue jay, no different than a crow. Because they will do the same thing, bother other species. I don't look at them as mean birds.

DENNIS: But do the birds that he's hassling consider him as a not-so-nice guy?

JOHN: Well, remember what Mike said the other day. We've been watching the crows and others. And we see the crows chasing the hawks and chasing the owls. And then the other day we were riding somewhere, and Mike said, "Look at that, now it's the crows' turn," because all the songbirds were chasing the crows away. I'm looking at that, and I'm thinking each bird has their own day where they're king.

MIKE: There's another saying, "What goes around, comes around." That's very applicable to Nature, though humans don't seem to get that.

DENNIS: John, are there certain species of birds that you feel particularly aligned with?

JOHN: Oh, yes. . . . Owls are the first ones. Owls have been connected to my family apparently for a long time. When my first brother was dying in downtown Toronto, where my grandmother lived, there was an owl outside the window for three nights until he finally passed. And my mother would tell me things about owls. I am not going to say what they were, but she would tell me things about owls. And I know when I was eleven my uncle gave me this mounted great horned owl. So, the owl is one of the ultimate ones. Then there are the hawks and the herons.

Each one of us has so many helpers. These helpers are aides to us. Why don't people look at them that way? Why don't they call on them when they need them and say, "Listen, I really need your help." The more you connect, the more you understand what their purpose is for you. And how you are so very much like your helper. It is really something.

DENNIS: Mike, you worked with hawks at Hawk Creek Rehabilitation Center?

MIKE: My primary job was to work with the volunteers to teach them how to work with, how to interact with, the animals. To set up parameters so that the people always pay attention to the birds and other animals.

Each one of these animals has their own gift. They have their own things that you have to respect. On the same hand, those animals have to respect that they have to allow the humans into their enclosures. Most of the people didn't get this. Because you have to work that deal out with each animal in their enclosure. If you don't work the deal out and say, "Look I'm not coming in here to disturb you. I'm going to respect the things that you need respected. But I have a job to do and if you understand," which they do, "I have to go in and clean and feed you." The birds eventually start to see. It's the mannerisms. It's how you walk in and what you are going in there with. They know. "Oh, he's coming in with food" or "he's coming in to clean." But you tell the people to always be aware when they're going in. . . . Do animals have bad days? Yes. And you better not go in there when they're having a bad day because you're going to encounter things, and you're going to say what the hell did I do?

Did I get bit? Yes, I did.

John, who has a special way with birds and other animals, says Mike has a very special way with animals. John had observed Mike get very close to a young goldfinch on a road, other birds, and to pond frogs who didn't hop away from him:

JOHN: Animals do not act normal with Mike. I am not being a smart ass, they really don't. They look at him as no threat whatsoever. So, when he says things about animals, that they are acting a certain way, I have to look at it: Is it because of him or because of the animals? The animals' behavior all has to do with Mike's spirituality. I think that's all they see. I don't think they see a human form when he comes up. I know that sounds weird. It's just unbelievable.

MIKE: They've done studies on how birds of prey get along with particular persons. And there have been veterinarians and researchers who planned to perform medical procedures on a bird of prey, but who arranged for someone else to do the procedure because they knew that bird would never be friends with that particular doctor.

One veterinarian taught his son what he needed to do to a bird, to replace him, and stood by to advise his son. But even the son could never go in the enclosure with the bird that his father previously, with much difficulty, had done. When you violate an animal's trust, it's for life. I would tell the volunteers that if you violate any of these birds or animals of prey, if you violate their space or that respect that you need to give them once, you're done. You only get one chance. So if you're having a bad day, don't go in there. There was so much to teach and train with the volunteers. And they're not all kids, there were a lot of adults. If you're having a bad hair day, don't go in there. It's that sensitive.

JOHN: I've been bit by just about every animal. Because how else you going to learn? Not that I enjoyed being bit, but you have to learn how to respect them.

MIKE: Because it's a new connection. When you start saying boy I got them mad, or I did something which led them to defend themselves—that's their only means of defense. We have to see things from their point of view. If we don't, if we're only looking from the human side, then we tranquilize them, we use force. Sometimes that is necessary. But I think the better result occurs when you work with animals to assist them, as John does, to provide help to them for survival. To do that there has to be a shift. You just don't think about your own life.

DENNIS: Something's leading him or you to want to touch the salamander or whatever. Most people would say I'm not going to touch it cause I'm afraid of getting bit. You're a little afraid of getting bit, but it doesn't stop you from touching the salamander.

JOHN: The salamander is a living–breathing thing and here it is, it's brand new. I want to learn as much as possible. It's no different than the time when I climbed the tree where a great horned owl was nesting. I wanted to go up. I wanted to see her, I got halfway up, and something told me to turn around. I see the mother coming toward me. I said man, I'm as good as dead cause I was up high. I was only twelve or thirteen. That owl flew by me, and I was waiting for her to nail me, but she flew right by me and only the wing smacked my back. "Get out of here, what are you doing?" she seemed to say. Man, I shimmied back down the tree.

With all these incidents in Nature, you get to realize that you don't understand then, but you do afterwards. After you become much more mature, I guess. You start to look at everything and you say, "Wow!!"

MIKE: It's the highest learning process.

DENNIS: Somethings sacred have happened, have touched you.

JOHN: I guess because of not having much attention at home, I was able to get attention from what was out there. You go and pick up a snake, and the snake bites you, OK. You are still hanging onto it, then all of a sudden, the snake becomes calm, it curls right up, and you go, man, this is pretty neat. But yet you can't go home and get help from your mother, so this becomes very sacred to you. This becomes very, very sacred. It becomes part of you. Then, another time, you pick up a salamander. Yes, it bites you, but it doesn't matter. I still do this to this day. I've been bit by almost everything. It's all right there. All that energy. You don't even know how it was made, where it was made, why it was made. Everything's right there. You can't get that out of books.

In the open sun, the phoebe perching was alit in gray-olive, almost yellow. He perched so naturally confident on the top rail of the weathered, wood place at his place.

The phoebe had four short whistle notes of acknowledgment that I was talking to him. As I was leaving, I felt guilty, because he might have thought that we were still in a conversation. I had put more than a moment into trying to connect, see, and talk to him, and then he acknowledged me. I console myself in part by my realizing late that my spatial leaving of him did not, from his perspective, need be a definitive goodbye, because he easily, to maintain connection, quickly could have flown ahead of me to reconnoiter yards forward. (I would be unable to do that—I CAN'T FLY!) But I still feel a little bad.

I told John and Mike about other experiences with phoebes, including two different times I found a dead phoebe.

DENNIS: So it was sad when I found the dead phoebe. He was my friend every year. Should I smudge or have some kind of ceremony?

JOHN: I put tobacco down to release the spirit. To give thanks for the bird giving up his or her life for you. When I go to skin the bird, if I am going to taxidermy it, then I'll smudge him. I'll have sage burning as I'm opening it up.

Do you have phoebe nests around your house, on the side of your house?

DENNIS: Yes. But I do not think there is one now because I think he, or she, is not around.

JOHN: I think it's he.

DENNIS: I think he only had arrived from his winter home a couple days before. I presume he got caught by the deep freeze here. I know they're pretty hardy, but it was a very quick and

very deep freeze. . . . I am sure you also must have particular
birds that come here every year, and you recognize them.

JOHN: Yes.

DENNIS: I read that the phoebe can come back for up to ten or
twelve years. Some other species and particular birds, like the
yellow warblers and the cedar waxwings, are consistent year
after year after year. The kingfisher that frequents the stream
next to my house is back.

*The phoebe came to see me. He landed on a mostly bare branch. I could
see his left eye looking at me or toward me. I saw the shine, the twinkle
of his eye. He obviously knew I was here before perching directly and
openly in my line of sight. Earlier in the day, our paths had crossed.
It's almost as if he missed me and wanted to hang out more together.
More than occasionally, he pops up in the trees right next to the house.
I know he lives quite nearby—on the grass lawn I once found a baby
fallen out of the nest, and I tried to raise her. She perched on a pencil
above a bowl on the kitchen table for a few days. I remember it well.
She died. I cried.*

*On a short walk, I realized, after seeing the familiar phoebe in
an unfamiliar locale distant from the house, that a cedar waxwing
had presaged the phoebe's unusual appearance there. I had gone to the
woodland edge to see the waxwing flock where they often appear around
sunset. But, with disappointment, I saw only one, briefly, initially. But I
now understand that the semicircle flight of the waxwing was actually
a heralding of the performance of an excellent viewing and mutual
cavorting with the phoebe. And then, just a minute after the phoebe
departed, a boisterous, handsome brown house wren flew waist high (to
humans) across the shrubs directly in front of me, where the waxwing
and phoebe had been. The three different species collaborated in a bird
show for my benefit. Perhaps their benefit also.*

*I had a triangle of bird friends—phoebe, wren, cedar waxwing—
all appearing within a couple minutes as I stood in the field bordering
woodland.*

EASTERN PHOEBE

Illustration by Julie Zickefoose

The phoebe (*Sayornis phoebe*), a dusky olive-brown bird, has a rounded crown, belly, and back with a fairly long, dark, square-ended tail. It is dark capped with muted white and yellow feathers on its underside. It is a flycatcher, a type of bird, and characteristically dashes back and forth off perches, five to six feet off the ground, where it patiently waits to capture flies and other insects midair. Generally solitary, it often returns to the same spot on the branch from which it took off. With exceptional eyesight and maneuverability, phoebes are expert fliers, and generally seen to have an "upbeat" disposition.

Phoebes have a short, high, oft-repeated call, *pewit-phoebe, pewit-phoebe*, and frequently dip their tails up and down. They frequently turn their heads right and left to patrol the flying insect air space and sometimes vigorously wag their tails after catching an insect.

The phoebe likes ledges and often builds nests in niches on houses, barns, and other rural structures. For a migrant songbird, especially a flycatcher, it arrives early and thus is a harbinger of spring. It also leaves late in autumn. Since phoebes often nest in human-built structures and are friendly, they commonly hang around occupied homes, and thus, for me, act like outside housemate companions.

THE CONTEXT: PHYSICAL AND METAPHYSICAL

John's place has a large workshed with his own hand-carved wooden bird sculptures in the window, a bird feeder observable from the kitchen window next to the table where we talked, and a large backyard that borders woods and a stream. In the back, there are facilities for raising and rehabilitating animals, including several medium- and large-size cages. In the basement of the house is a turtle rehabilitation center and turtle raising laboratory. By the backside of the house, not visible from the road or front of the house, is a sweat lodge.

For John, the purpose of the sweat lodge ceremony is to allow insight and revelation to occur. To provide an environment where people can help transform themselves into better human beings. As mentioned previously, John was trained in the use of sweat lodges by Seneca Chief Skye. Sweat lodges have long been used for purification ceremonies.

Indeed, one might argue that John's "mission" is to improve the lives and well-being of humans and animals, humans *in* nature, in alignment with Nature and its Natural Law. Everything is geared toward Natural Law in alignment with Native prophecy, and to offset the growing environmental crises. His work honors the Creator's works and eases the likely upcoming dramatic earth changes brought on by humans' ignoring and disrespecting animals, each other, and Natural Law. Natural Law, the Creator's program for the planet's harmony and sustainability, is the model that humans should follow. Cooperation and respect within all elements of Nature. (See the conclusion and afterthoughts.[2])

JOHN: I've always wondered, back when we started the sweat lodge twenty-two years ago, over those years how the teachings have been to improve the people who come into the lodge. It's always been to improve themselves and at the same time to learn lessons being taught, whether it's the animals bringing the lessons or whether it's Spirit. I can see that people who didn't pay attention or just blew off the lessons are having those lessons really show up on them now. It shows up so strongly with them. It's like whatever you didn't get rid of you are going to carry much worse.

John's approach accords with others' traditional use of sweat lodges. Not only are they used for purification or to clarify thinking, but also for small groups or community help and community building. Sweat lodges are also built to (re)connect what he calls the Three Worlds of the Spirit side, the Universe side, and the Earth Mother side.

While talking around the kitchen table, the three of us occasionally took note of the birds outside the windows: chickadees, blue jays, crows, nuthatches, juncos, and others. Sometimes we heard bird calls and songs from near and afar, especially as the calls appeared as comments on things we said.

One time around the table, as Mike was talking about animals and humans, he heard a blue jay, and he translated his/her supportive message:

"See. Now." That blue jay is saying, "Yeah, well, why not, how come the other [non-Native] doctrines do not include the environment. What have the humans done to our home!"

JOHN: This is the way I believe it. I don't know about Mike, but this is what I've learned from Nature. It always shows where the problems are. Whether it's an individual or whether it's a whole species, or a man or woman.

DENNIS: So, it was on your own that you picked up the ability to draw and carve birds, and to read bird signs and deer antlers?

JOHN: Well, only in part. Chief Skye had taught me about the meanings and use of the right and left side of animals. I think it was in a lodge. In the second lodge I went to, they talked about how they were having a problem with this one side, and after the lodge everybody was talking, and that is when I found out about the right side and left side. That's how I learned. And then once you find out with practice, you see that it really works. For example, if the bird has a broken right wing, I have to look at my male side: Am I being over-dominant on my male side?

MIKE: Besides drawing birds, John has another gift in his taxidermy. John honors the representation of the birds. They are not simply animals. They are life-forms which have a higher purpose. Birds need to be honored, because they are still living the life that they are supposed to. To see a bird or other animal taxidermied, it is not for a trophy. Here is a gift of their lives. They gave their lives, being represented in their form. They gave their lives so that their representation of who they are is still visible, in the physical. They gave something to be here. Whether or not they are alive, they are still physical ambassadors for those animals, to be treated with the same respect as living ambassadors. That is key to what most people do not understand about taxidermy.

JOHN: When Marsha [pseudonym] moved in eight years ago, she hated taxidermy. The minute she saw a mounted animal, she hated it. She thought I was like all the other taxidermists. I said no. You have to learn from the animal. You are doing a service, but are getting the lessons as the animal comes in. It is really different than what most guys are doing when they preserve bird or animal specimens.

MIKE: Most people do not connect any other elements, other than that they shot something. It really is a higher representation of an animal. The taxidermized birds have sacred connections.

DENNIS: So they really are sacred objects.

MIKE: These stuffed forms shed more light not only on the birds, but all those life forms that are here. John's practice is a much higher medicine than taxidermy. I do not think many people understand this. Actually, it is the wrong word. Taxidermy should not be used to represent animals that have given their lives. They are still here to show that we want to honor who they are through the ability of preserving the physical elements they had. I guess by dictionary definition, that is taxi-dermy, but it is not the right word.

DENNIS: I guess the stuffed birds provide a kind of life force or sacredness that remains after they are dead, and humans can pick up on that energy. Live birds, of course, pick up on the energy of humans.

JOHN: That's why I really like having here in the kitchen the screech owl that I am trying to heal. Because that owl tells me about visitors coming to the door or outside the windows. The screech owl acted up before someone even came. She did when a certain person that the owl didn't care for came here into the kitchen. . . . It depends also on how that person acts. People carry stuff and so if that person was going to be what he normally was like, which was disturbing, the screech owl would carry on and I would ask the person to leave. But if that person was in a rare mood in which he was being a good human, I would let him stay. But I would always watch the owl, as the owl would watch the visitor closely. . . . Look at him right now, with the three of us. He's nice and calm, but sometimes, in the beginning, he acted up and went bonkers because he didn't know you guys.

DENNIS: Have you seen that birds actually enjoy certain activities, particularly flying?

JOHN: Yes, with rehabilitating birds we've sometimes had a gust of wind that comes up and you'll see them just open up. I believe that they enjoy it. I really believe that they enjoy flying around.

DENNIS: Yes, I can imagine that. I know from my experience with chickens that I can tell a happy chicken from a not-so-happy or sad chicken. We've had chickens for many years. Some of them are full of anxiety, usually the hens or non-dominant roosters. It's like you can tell a happy horse if you look in the eyes of a horse. I can see it in a chicken, but I don't know the wild birds well enough to say whether a particular bird of the group is enjoying him or herself or not. Certainly chickadees, for example, with their bright eyes and actions frequently seem like they are happy.

It had stopped snowing two or three hours earlier. Now the full sun shone in an all-blue sky and reflected glorious amounts of light off the white snow, having accumulated to over two and a half feet. On two wild apple trees, four black-capped chickadees merrily chirped as they flitted back and forth among the branches. Mostly from branch to branch pecking at the few still-hanging, now-brown apples. The apples must have been frozen, except for the outer parts thawed by the bright sun. It was a happy event for all, the birds, the trees, and whatever gods were smiling upon the picture-book winter scene.

MIKE: Birds, because of their design, the way they are created, are not allowed to show to other species that they're in trouble. Because as soon as they show signs of that, they're prey, they're food. That's why some birds, like the killdeer, will make a mock injury in the presence of humans. The only reason they'll do that is because they know that the humans will follow them. And it works. Now think about that, how did they even come up with behavior like that? They don't have books that birds read that say if a human gets too close to your nest you hop down on the ground and hold your wing funny, and then hop sideways and they'll follow you. As a kid I did frequently disturb some birds. But then, when you get too close, they take off to draw you away from where the nest is. That's really cool.

DENNIS: So, if a bird were to show emotion, it might mean s/he would end up as somebody else's dinner, because s/he is no longer in control. Emotions in humans we know often get in the way of clear thinking and focus.

MIKE: The bird wants to show no symptom, nothing, that would be an indicator to the life around them that they have a problem. It doesn't take long for the rest of that life around him to go, "Oh, there's an easy meal." . . . But the birds know. Like when I was working with that goldfinch because he had pink eye. He couldn't see. But they even know if they can't see. They can still sense. He knew when I saw that it needed help: I just picked him up, put him in a box, and brought him over here to John's. They know how humans can't seem to hide their intent. Humans can't keep that energy from being projected out. So the birds do pick up on it and they do recognize the humans who are "OK." But most of the humans aren't. So the birds are fairly confident and safe in not trusting the humans. But even as John experienced with the seagull, the bird didn't resist, didn't give him a hard time, when he first picked him up, because he was in trouble, and he knew he was in trouble. But he was in trouble in the sense that John was going to give him medical help, not in trouble in the sense that John wanted to eat him.

DENNIS: Almost always, most birds, as you guys know, are very hard to approach. They seem always very cautious. I read something in Charles Darwin's journal, just by accident. One of his journals describes his going to the Galápagos Islands for the first time. He said that when he got close to the birds, they would not fly away at all. In fact, he had a hard time because the birds were all around him, even in the way of his walking. (I recently talked to someone who had visited the Galápagos and they said, likewise, that the birds would stay so very close.) Darwin said if he wanted, he could club them over the head, they were all so close. They had no fear, I guess. No worry about the fact that humans were around because they never or rarely had seen humans before. Now, here, blue jays warn the other birds and animals when hunters are around with guns. It is an adaptation to the fact that they know that the humans here are not always, or maybe ever, nice to them. Right?

JOHN: I believe every animal knows. If the animal feels as though you are not connected or knows that you are not connected, s/he is just going to stay away. Because s/he looks at you as something that might harm him or her.

MIKE: Behavior does change when animals recognize that they're being taken, whether it's for food, feathers, fur, or whatever.

It changes the relationship with the human and the future of these animals. Deer here in western New York also know. Again and again, they hear gunshots being fired and they see the change of more and more people coming into their domain and habitat. It's interesting to see how Nature retains information passed down from previous generations. So each generation does not have to start over and over, not trusting the humans.

DENNIS: So, you would argue that they pass on the traits to one another. So that's a spiritual process, unless you would think it is genetic?

MIKE: Spiritual on one level. But I think most researchers would not want to bring spirituality into the concepts. They want to keep things as instinctual. I mean there's such a resistance within that community. I guess it's partly scientific, but they just seem to struggle with what influences behavior other than just physical contact. . . . And you know there are communications that constantly go on amongst the members of the natural world, and it seems to be that humans have distanced themselves from what's going on there, and because of that humans seem to not have a clue.

DENNIS: The birds pick up on the humans not knowing?

MIKE: Yes. And it's not just the birds. I think all of the life out there recognizes when humans have detached from the Natural Law.

DENNIS: So the animal or bird can detect in you or me or someone our attitude or our understanding. They just kind of know it and can communicate that. So adult bluebirds or blue jays teach the baby bluebirds or the baby blue jay to fear, or at least avoid, the human.

JOHN: Yes. Because the bird's babies watch what the parents do. So if the bird sees a two-legged walk up to the adults, and the adult bird takes off, and this constantly happens, she's bound to settle into the baby to say, "I have to flee, here comes a two-legged," or "Here comes a cat."

MIKE: If you want to do an evaluation of yourself, the best way is
to walk out into the woods. If everything takes off and doesn't
come back, you better look at yourself. Typically, the animals
know just with you walking into their territory that they are
going to get away from you. But if you show and demonstrate
why you are out there, they are going to come back. You will
see woodpeckers or chickadees or deer, they will come back
close to where you can observe them. Most people don't get
it. So that's the best way to get evaluated.

If you think you are not being watched when you go outside your door, then you are not paying attention.

DENNIS: There are certain species of birds that live in the city
or spend time in the city, although many of them can't or don't
want to live there. Are the birds that choose to live in the
city, often in very concrete areas, consciously trying to bring
sacredness to urban dwellers who don't have access to the
countryside's huge world of trees and other plants?

MIKE: Yes, they do. Some species and individuals go to the city
to bring lessons to humans, as they can adapt to that kind of
environment. Sparrows, doves, seagulls, and others. A lot of
the birds won't go there because there is little sacredness.
But a few years ago, I was on West Ferry [Buffalo]. I looked
off to the side and there was a red-tailed hawk sitting on a
sign, right in the city.

DENNIS: "What are you doing here brother?"

[Laughter]

MIKE: Yes, "This is not where you should be," but obviously that's
where he was. And even a friend of mine told me she's seen
birds that you typically shouldn't see in the city.

JOHN: Peregrine falcons are up in Buffalo, over at the old train
station.

MIKE: And lots of pigeons, seagulls, and crows in the city. They
like the high ledges and buildings. They are good nesting sites
for them, and there must be enough water and ducks nearby
to feed them. But birds aren't bound by man's laws and rules,

and they're where they need to be, and that's where they are. So we just thank them for being where they are, what they're doing, what they bring. Some of the humans will get it and some of them won't.

Birds have a life force, and they all contribute not only as individuals but at the same time they are part of a collective. That is why everyone does what they do at each level. Some are seed eaters, some are bug eaters, and some even eat each other. That is how the collective functions. Every element has been provided the food that they need, and they serve their purpose.

And with respect to birds such as seagulls (sometimes crows), often considered by Westerners as dirty, annoying scavengers:

MIKE: Seagulls are housekeepers. That's what their job is. Mad Bear used to say seagulls are what people call scavengers, but they eat things that no one else wants to touch. Mad Bear did a comparison between the crows and seagulls. . . . One is to clean up the waters and one is to clean up the land. Mad Bear said when you see the seagulls all coming into the land, that was another indication that the changes were very close. You can see them at night all going back to the water. I see them out by us, and they actually start forming Vs. The seagulls start flying like geese. It helps them fly back to the lake. I see them all migrating back to the lake about an hour and a half, two hours, before sunset. . . . Seagulls are pretty independent, and they want to get what they can get. Crows seem to be more cooperative with each other. But it's an interesting balance to see two types whose designation is to eat things that humans don't want to touch.

DENNIS: Yesterday I saw a kingbird on a telephone wire eating a large dragonfly. I saw the long body of the dragonfly because the kingbird ate it headfirst. I saw the dragonfly's wings as they went down the bird's beak and throat. It was neat. Although maybe not so much for the dragonfly. It is very busy out there in Nature when you stop to appreciate it. You know, I just go and sit in my backyard, I sometimes call it, the huge field behind my house, or in one of the fields. I just sit there and if you just sit there, there is so much going on. So much activity. You cannot hear it all, but you can feel it. It is amazing.

MIKE: It is like when John and I went to the lawyer's office to talk about the environmental issues. Even as we were driving up to the lawyer's office, crows were coming and landing in the trees next to the road. We saw many crows on the way. When we got there, I saw two crows sitting on the roof of the church next to his office. I was in the line of sight with them, and they flew to get out of the line of sight, but they flew closer to the lawyer's office, and they stayed in the tree next to his office. I could see them through the window in the reflection of the glassy surface on the table in the office. And you know, the crows were watching the whole time. They were interested: so we had better keep them informed. Well, in the discussion with the lawyer, we went as far as we could. The crows were still outside the window of his office, and they sat there for the whole consultation. And no one else could see them, but we saw them in the reflection. They sat for an hour, but normally crows will only sit for a little while. But they were watching as sentries. As soon as the lawyer asked the question about how far we wanted to take the legal case, and I said as far as we have to, the two crows took off and flew away.

Mike and John are here referring to John's meeting with an attorney, who defended John against the New York State Department of Environmental Conservation (DEC), to which he had earlier been a consultant. The charges against John were for saving and rehabilitating birds and other animals without permits. Although he was charged with numerous violations, which could have resulted in months in jail, the judge sentenced him to only the smallest possible fine. This was done in part because expert testimony indicated that his activities were based on religious grounds. Arguments indicated that his "religion" required that he take care of and rehabilitate injured birds and animals.

DENNIS: So, the crows knew, know, that you were working on their behalf?

MIKE: Yes. They were allies. I told John, "They are listening. They are not only listening, but they are going to tell all the other crows."

JOHN: Indeed.

MIKE: John was and is doing the right thing, and the lawyer knew it. And I said if we do not stand up for doing the right thing then what the hell are we doing here? I said this is the part of life that says that it is our obligation for future generations.... So those are the powerful parts of life that say that if we do not work with Nature then what are we doing? Is Nature just a resource to destroy and abuse? Or is this part of the life?

The crows, I think, because of them being social, are on a higher intelligence level than other birds and many humans. Social types of activity enhance behavior and that gets more recognizable to humans. There is something there that the crows seem to understand.

JOHN: Yes.

MIKE: That's all they wanted to hear. I think there are individuals in every species out there that have a higher level of function or interaction, as do some humans. Why wouldn't or couldn't that be true?

The crows are the best allies to have because they are very social, and they protect their families. That is why they are going after hawks. They are not going after hawks because they think they can eat them. They are going, "Man, the hawks eat other living things, they may even eat birds, we do not want them near our young." They are social and keep their young with them for a few years. Their family, their community, grows. I have seen fields where there are 200, 300, or 400 crows out there. And they council twice a day, once in the morning, and once in the evening.

Crows, hidden, calling back and forth from trees on each side of the overgrown pasture. Talking back to each other. After a few moments, the volume diminished. Then, four big, black crows cawed across the field.

DENNIS: So, in this scenario, which species do you think are a little more special in the ways that you're talking about?

MIKE: The crow is the first one that I think of as a "special" bird, but then I start to see those others that are related to the crow, like the blue jay. They both warn the deer, the squirrels, and all the animals during hunting season. They go out there and they see. They know and you wonder how do they know? Well because the humans developed a new behavior going out into the woods carrying this long thing [rifles], and they make the connection. The birds observe, they watch. . . . And when you start to see how they connect, they really are using that higher intelligence for their own protection. And they're not becoming aggressors, they're not turning reckless or trying to hide. They're just working with humans' behavior changes. It's showing us that we need to pay attention to what they're saying, what they're doing. The animals are here for our benefit, and not just because we eat them. They're here on other levels to interact with us.

DENNIS: The other day, after I was walking in the woods, I sat down on a log. After about five minutes, three crows flew out of a tree that was about twenty yards away. I hadn't seen them because the leaves were full. Obviously, they had been there since I came to that part of the woods and were there while I was sitting down. They must have known I was there.

MIKE: Sure.

DENNIS: How far away can I or anyone be for the crows to take notice? Or to keep or take new perches with respect to me, rather than just leading their own lives without regard to me or any other person nearby? They are not always messengers to humans, are they?

JOHN: But they are always watching us. They always have one crow someplace who is always watching while the others are feeding.

MIKE: But it is not out of fear. There's no fear out in nature, as John said the other day. It is because there is an intelligence that says "post someone to warn us" while we are busy hunting for food.

DENNIS: Do crows, more than other birds, have sentries?

MIKE: Yes, because their structure is more social. Hawks and birds of prey, even robins, catbirds, and woodpeckers do not group and socialize like the crows. I noticed this phenomenon right after 9/11.

When there was more patriotism and invocation and use of eagles. I said why from a human perspective is there an attraction to birds of prey? Eagle or hawk, humans seem to look up to them. We have a high opinion of these birds of prey and yet they are not social. They keep very much to themselves and do not interact well with other birds. Other birds do not like having these birds or owls around.

One day, trying to read bird signs regarding getting together with Mike and John, I spoke about a woodpecker.

DENNIS: That day last week when I thought we were getting together but couldn't, I went out to my barn where I had parked my car. On the tree right in front of the barn there was a downy woodpecker, a female: it didn't have any red on it. The downy and hairy woodpeckers are often around the land where I live, but I had never seen or heard the downy woodpecker right around there or on that tree. So, I wondered what was going on. For that day I had prepared in my mind to ask you two some specific questions about birds, about particular birds, the crows. . . . What do you think the woodpecker was chirping about at me?

I thought maybe the woodpecker was saying something like "don't forget about me, you bastard."

I had had with me non-Native Ted Andrews's book *Animal-Speak: The Spiritual and Magical Powers of Creatures Great and Small*. Mike suggested that we see what he said. I read out loud:

> Sometimes the woodpecker will show up just to stimulate new rhythms. . . . Sometimes it is easy to get so wrapped up in our daily mental and spiritual activities that we neglect the physical. This can be when the woodpecker shows up. It may also reflect a need to drum some new changes and rhythms into your life.[3]

DENNIS: Do you have your own ideas about the downy woodpecker?

JOHN: Well, it is about the same as that. Whenever we see woodpeckers, that's the kind of message I get, except for the flicker. The flicker is more of a medicine bird, to me, anyhow. So, s/he is usually telling us to listen to the bird's heartbeat, which means "stop what you are doing and get back connected again." That is usually what it means. . . . We get too hung up with all the world shenanigans.

DENNIS: So, was the woodpecker there for me before I saw her and heard her? Or for whoever was coming by. Or was it a coincidence that she just happened to be there when I was going by?

JOHN: I don't believe in coincidence.

MIKE: Coincidences fall into an area between science and superstition. When an individual refuses to see or denies that things happen for us, and to us, for a reason, sometimes people cannot figure the reason out. But what's going on is that it is just one more lesson that says, "Don't let things keep going so long, do not avoid addressing it. Stop it and fix it before it gets to a dysfunctional point." So you know there are reasons for everything we experience, whether it is the woodpecker or something else. Those species know by a number of energies. They can detect the frequencies. Everything can communicate and everything can read. Humans actually should have the highest ability to read these things, but they gradually keep dumbing themselves down with technology and greed. . . . The success in life is when you see beyond the stuff which clouds your vision. You say wait a minute, what are the woodpeckers saying, what are the crows saying?

The Creator said every lesson, every teaching, you ever need in your life is right there, right here. It is not in a book; it is not written in poetry. It is stated clearly in front of you every day. It is in the trees, it is in the birds, it is in the four-leggeds, it's in the fish, it's in the turtles, it's in every level of life.

DENNIS: John, you were saying that the flicker is more of a medicine bird than other woodpeckers?

JOHN: That is the way I look at it. . . . To me flickers are the ones that actually create the thunder. OK.

Flickers are very strong medicine because they carry so many different colors and different things. I have heard that many medicine men will put a flicker feather in their hair. They are a bird that is totally different from the other woodpeckers.

DENNIS: Were you shown this meaning, or did you learn this from observing the flicker?

JOHN: It was through a vision, that I had out at the [sweat] lodge. . . . Yes, we had a couple of flickers that were hit by a car and had come to us from people who dropped them off. So, when we had a lodge, I offered up for the flickers. I gave thanks for them coming. And so, after that there was a lightning bolt. And on the lightning bolt was the flicker. Actually, there were four of them and all the lightning bolts came together. When they came together, it made this drumming sound, and it turned out to be thunder after the drum sounds. It became thunder, and that is about all I am going to say about that.

DENNIS: When you say the flicker is a more powerful bird, you mean the flicker is more spiritually powerful because they have more medicine?[4]

NORTHERN FLICKER

Illustration by Julie Zickefoose

There are a few subspecies of flicker (*Colaptes auratus*). In northeast North America, the usual one is called a northern, a yellow-shafted, or a common flicker. It is the only brown-backed woodpecker, and the only one who commonly feeds on the ground, although it hops and runs a bit awkwardly. It likes ants and catches them with its long sticky tongue. Like all woodpeckers, it climbs up and down trees, using a chisel-shaped beak to peck, and stiff tail feathers to cling upright, two toes in front and two toes behind to help hold onto tree bark. The flicker uses its beak less than other woodpeckers but digs high holes for nests and winter homes in decaying trees. In spring, it beats a loud tattoo on dead branches or tree trunks, sometimes a telephone pole or roof.

Its rather complex coloration has feathers of several different colors and patterns: black stripes on the brown back, yellow under the wings and tail, a white rump, a black crescent collar across the breast, and big black spots on its light milky chocolate chest, and, in males, wide black whiskers. It also has a red patch on the nape. Altogether, it is an eye-catching bird that undulates and displays a large white rump patch as it flies. It is found in different habitats, from city parks to suburban lawn to remote wilderness.

Nelson relates that for the Indigenous Koyukon of Alaska, the flicker "is a source of great power, capable of protecting people and bringing them a lifetime of good luck. . . . Someone who hears flickers more often than normal should have consistently good fortune."*

There are many common names for the flicker, including yellow hammer, high hole, yocker-bird, harry wicket, yarup, wake-up, and pigeon or gaffer woodpecker.

* Nelson (1983: 112).

JOHN: I do not use the word powerful: The only thing that is powerful is a thunderstorm or waterfalls or things like that. I don't ever say that animals are powerful because I believe that they have different strengths to them for our own teachings, lessons, or medicines. Each form of wildlife has their own strength and has their own special purpose.

DENNIS: Does that mean that in winter, when fewer species and birds are around here that there are fewer lessons or messages available?

JOHN: I believe that in the wintry time of the year we should be looking at ourselves, and the birds basically tell us about ourselves. For example, the blue jays, they're like royalty. Are they like you, looking at yourself as being some important thing? Or are you looking at yourself like the juncos carrying all that weight on top, what with their dark backs on top of light breasts. Plus, the chickadees bring the truth and nuthatches tell you to keep the faith. They're like our miniature faith keepers. Then you have the woodpeckers that tell us to keep focused on the earth mother. So, you know, if you put all that in this time of year, and really look at it, it makes total sense. That's the way I look at it.

DENNIS: So, are you saying that the messages, or the potential lessons, are more concentrated in fall and winter because there are fewer species around then? Since species like goldfinches, waxwings, warblers, etc. no longer distract you from your inner self?

JOHN: The goldfinch now [autumn] has changed into a dull green color. He looks almost like a sparrow, so you know that brandnew beginning color isn't there anymore. I've always looked at wintertime or falltime as when everything starts to sleep and it's time when we start to look at ourselves and see where we've come in one full year. To see where we're at. Once you really stop to look at yourself, we can see how we can better ourselves, and start to see where we're headed.

MIKE: The information that the birds are bringing is the same message. It's just in you and the change of the season.

DENNIS: But how about in the spring? As you were saying, winter is more of a time for inner understanding, working with the birds to take stock of oneself. In the springtime, we get these various species from other parts of the world. Ironically, many of these birds are here for three or four months and spend more time elsewhere, often in distant, foreign lands.

JOHN: Right.

DENNIS: What is the general program [laughter], in the springtime, what with all the orioles and the colorful birds? The rose breasted grosbeak, the warblers, the bluebirds. They have all these colors coming. Are you saying that it's not as much as a time for inner searching or working?

JOHN: I think you know what birds are going to come back at different times during the spring. I think if a person looks at their life and looks at the meaning of the bird, I think they're going to find that it's very fitting. . . . I've never seen it be otherwise. I've birdwatched all my life. And more and more recognized what birds mean to me. I know that when the orioles come back, that's a special time for me. When the red-winged blackbirds come back, that's a special time. . . .[5]

Above and across the field a red-winged blackbird chased a brown thrasher until he landed a couple times on a big honeysuckle shrub by the edge of the field. Then the blackbird flew back along the previous flight path.

JOHN: And the list of colorful birds goes on and on. They boost you. I think they boost you when you see the colors at different times, so that you don't get complacent and just say, OK, that's a red-winged blackbird. You suddenly see a bright orange and black and you're going, Wow! That's a Baltimore oriole, and you seem to get a sense of newness, like fresh air.

DENNIS: Right. It's happy, it's refreshing.

MIKE: Well, if we appreciate that change in the color. Even with the autumn leaves, I think it's another representation.

JOHN: Sure, it is.

MIKE: Of saying that all that life works in a harmonious way. It's not just the birds that have good color. Look at what the leaves do. . . . There was a lady at work who came up to me last week. She described a big owl sitting in her parking spot at home. She pulled in at night and she said there was a big owl. The owl flew up and sat in a tree right near where she parked. I asked, was the bird a great horned owl? She said no, the owl didn't have ear tufts. She looked the owl up online and she said the bird was a barred owl. I go, Wow. . . . I said to her that she was lucky. Rarely do you ever get to see them in the wild.

JOHN: Is she going to pay attention to the owl's appearance?

MIKE: What she read was to look inward and try to work things out with a big problem she's struggling with.

JOHN: The barred owl stands for the need to make a choice. You have to make a choice as to what's going on. There's a fork in the road and you had better realize it.[6]

MIKE: These are part of the lessons.

DENNIS: Did the barred owl know that this woman might pay attention to the lesson the bird was bringing?

MIKE: The owl knew. But I mean not all humans know. Not everybody would even think about looking up or talking to somebody about the appearance of an owl, or any bird. They'd just say, "Oh beautiful owl, or spooky owl, nice to see you, goodbye."

DENNIS: So how does it work? There must be people, including me, who are on the border in some sense that we don't quite know or understand these things, these birds, these lessons, like you guys do. I understand part of it, but I don't see it as much as you do, I don't feel it as much because I have a different background. There are times that I can imagine for other people that they may or may not feel a sense of connection or even think about the possibility that the bird is there for a reason for them. How does the owl know that maybe this person might listen to it?

MIKE: I think it's more than a maybe. The natural world just has a knowing. There's really no explanation for it. There are no typical words that might describe it. A lot of times there are no words in the Native language for a coincidence. You don't want to say it's intentional, but it's collective. It's a knowing we already know. We're here to help the humans. We're here to bring them a message if the human stays on track on a level that says maybe I should start paying attention. John points it out to me all the time, as a hawk or other bird shows up, like when a red-tailed hawk was sitting right in the middle of the street by us.

DENNIS: That happened to me a few days ago, with a friend of mine. We were talking about birds and there were no birds around that we could see. We just happened to turn our heads back and there was a red-tailed hawk. If we had not turned our heads at just that moment, we would have missed it. But it had to be. For a few seconds, the red-tail was there to tell my friend that I knew what I was talking about. The bird's appearance was a message.

MIKE: On one level, it's confirmation. You get reassured that what you're working on is the right stuff to be doing.

JOHN: And it makes you look at yourself and say, "Yeah, I have to look at myself."

DENNIS: Sometimes I get attacked by the barn swallow. Does that ever happen to you, John?

JOHN: Oh yes.

MIKE: Well, it's interesting you say that because people who owned the large cabin where we were staying put up little platforms for the swallows to build nests on. But the swallows will not use them. They want to build their nests the way they want to, they don't want a platform under it. So, they're still attaching their nests to the side of the stone. It shows me that the swallows want to maintain their own higher function. . . . Maybe this is not the best analogy, but there are certain humans who will sit there with their hand out. . . . They just want something put in their hand all the time. They don't want to take the initiative. I think it's more than self-respect, it's living the image that you want portrayed in life. You don't want to be known as the one who likes to just live on the edge, and somebody will make life easy for you and they'll take it.

So I think there are so many more insights here. That's the other part that interests me. These insights, how the animals have them. The animals can show us. At the cabin now, they have probably twenty-five or thirty swallows and in all the areas around that lodge not one ever dive-bombed any of the people. I think it's because the swallows had enough human connection, and that's when the birds realize the relationship is sound. And the humans and the birds keep busy talking to themselves. They can keep doing what they were doing. No need to attack the humans to get their attention.

DENNIS: Earlier you mentioned somebody carrying owl medicine. I'm not sure what you mean by that.

JOHN: Well, we have our own helpers. Like with me, the owl has always been there ever since I can remember. When we were able to get a live owl, we did, and spent twelve years with the owl, teaching us. I believe that you pick up and start carrying that. I asked Mike the question, do the animals sense that you're carrying medicine? Will they shy away from you because owls are predators?

DENNIS: So that some of the owl, as predator, is part of you? So the animals or birds that are the owl's prey, or that are afraid of owls, might pick up on the fact that you are a predator person, or a predator owl person, or something like that?

JOHN: Yes. But you can even go farther with it. Great horned owls like their little areas and that's where they're comfortable. That's where they want to stay. Someone with different types of hawk medicine actually like to go long distances.

I like where I'm at. So you start to realize the animals you're connected to start to teach you, and that you start to take on, or already have, those qualities.[7]

MIKE: You do pick up animal traits in varying degrees. I think you can pick it up if you're living with an owl or other animal. When they're in your space, you share that element. After a while, the owl gets comfortable, and she knows she's in a safe place and can start releasing or sharing other elements that normally the owl wouldn't do with a human because they usually don't have the close proximity.

JOHN: I had the great horned owl for twelve years. . . . A person called me up and said they had an injured crow. I went over there to get the crow, and that crow bit me like you would not believe. It wouldn't bite my wife, but it bit the crap out of me. Every time I looked at this crow, he started going crazy. Luckily it was OK: two days later we turned the crow free. But, oh my word, that thing was horrible on me. I was going, "Why don't you like me? I'm trying to help you, you like everybody else, but you don't like me." That's when I started thinking about how crows don't like owls. One of the crow's jobs, according to the owl story, is that crows are supposed to keep the owl hidden away during the daytime so that the owl doesn't come out. The owl is only supposed to come out during nighttime. That's part of the old story about the owl. That's one of the things, when you start to look at your own animal helper, you should start to learn the old stories about them—how they got their eyes or their colors or this or that. It's really all part of it. This way you know from the old stories how it came to be and you're learning every bit of it.

DENNIS: So how do you access the old stories?

JOHN: Most of them are told by elders and others. Some have been written down in old books, and, nowadays, some of them are also on the Internet. There's one man who does a lot of stories. When I was interested in how the owl got its eyes, he talked about the whole story. It's really a neat story. Then you hear some too from Chief Skye and other Native people who will tell you different ones.[8]

JOHN: Let me ask you something, Mike. If someone you know has always worked with owls, does that someone carry that owl medicine such that birds recognize that on a person? So that would make the bird leerier of being around that person or want to attack them?

MIKE: I think there is a connection there.

JOHN: I think so too.

MIKE: It's not as clear as some of the other elements. Ultimately, humans can't be trusted. Because we have the ability to mask or pretend what we really are.

When we move into the spirit helper element, I think that having a spirit helper calms the animals' responses because it's not the human. In the Spirit element you can't pretend.

You can't fake spiritual connection.

JOHN: Right.

I heard a call, mistakenly thinking the bird was a crow, when, after about five seconds, I again heard the call, like a quack. At the same time, through the opening in the sumac tree canopy above me, I saw fifteen feet above a pair of great blue herons flapping their large wings.

Another time John spoke about a dead, great blue heron somebody brought to his house:

DENNIS: When he brought the heron over, did she stimulate anything in you about you and herons, or about the story of herons you learned?

JOHN: This one had food in her stomach, so it was different than the other. I'm going, OK, she has food, but I've never seen one with a neck that was damaged the way this heron was. After that I started looking for things that would show why she came. All I kept getting was something about a lack of patience.

DENNIS: You said that you also received two dead red-tailed hawks, which still had dead moles in their claws. That's unusual that they hadn't let them go.

JOHN: That's unusual. Usually when something gets hit or killed, what happens is that the bird's going to release it. So, it still had strength in its muscles. And both were moles in their claws, not mice. Chief Skye told me the way we believe is that the moles mean that something's trying to sneak by you. The mole was in the hawk's left foot. Males are usually associated with the right, females with the left, plus you also must look at the person who brought it. The woman who brought the hawk is the one who brought both hawks with moles in their claws.

MIKE: Was this Alice [pseudonym] who brought you the hawks?

JOHN: Yes. So, you know, it's kind of fitting.

MIKE: I'm seeing a lot of hawks get hit this year.

JOHN: One of them was a young one, but the important part was that the hawk still had a mole in the left foot talons. I just looked at her when she brought the red-tail over and I said, "Well that's number two for you." She went home. She had a dream. The dream was that she was walking a distance away from and parallel to a bear that was walking. Parallel, but they weren't together. I asked her, "Why weren't you walking with the bear? You should be walking together, not having a gap in between you." Finally, I said, "Does that ring a bell with you?" Because she wants to walk the path really bad, but she's not really willing.

She just looked at me, and I said, "You had two hawks that came and had food right in their hands." They had all that nourishment but never ate it. I said, "Well maybe it's something you need to look at with your past. Maybe you're not looking at the nourishing part, the part you should be looking at." That there, to me, was an important teaching.

JOHN: One day Alice brought over this man. The night before I had received a bird with a right wing out of joint. That's the male side. I said to myself that this bird's situation is going to fit, to parallel, this guy's situation perfectly. She wanted me to talk to this guy. He finally says that Alice came to rescue him. I said, "You have to rescue yourself. You can't wait for someone to come and save you. You have to make the effort to rescue yourself—you can't dump that on another human being." . . . Same thing with this bird, you don't want to fly? You're going to die, so at least give it an attempt. Let's see what you can do. . . . Sure enough, the bird did get better and better.

Another time, the three of us talked about bird and human travel and migration.

JOHN: As for red-tailed hawks sticking around all year round, there are a number of birds that stick around. I look at them like I have heard about the caribou—they always take the same path to and fro. It appears as though they are constantly going through the same lessons, crossing the same river, at maybe the worst time. A lot of them die off or are harvested by a pack of wolves. Caribou are constantly living the same lesson. . . . I had a friend who brought over a caribou skin. I looked at it and I asked, "Do you do a lot of moving?" And she said, as a matter of fact we are getting ready to move out west again. This woman would stay maybe a month or two in one place and then move again. I looked at her and the words just came out, "So you never got the lessons, you have never learned them." I felt bad because it just blurted right out, and she wanted to talk about it, and I said maybe she needed to look at it. Maybe you need to look at how moving does not help your situation. You need to find out why you have to move all the time.

I wonder that, like with the owls that usually stay in their own little spaces. I am a little like them, I'm not much of a traveler. I don't like to travel, I'll go little distances, but that's about it, long distance no way. The same thing with the turtles, the turtles are small, and don't travel far.

MIKE: I used to like to travel. Now I don't like it as much now. But I would always come back to the place where you call home, and I think that's one of the keys. To say we have to put our roots in place. Once you establish what you may refer to as your roots you come back to the place because you're familiar with it, you had good experience there. It works, and it's part of the life.

DENNIS: You belong there.

MIKE: I do know people who are constantly on the move. Every few years they move. They seem to think, "Oh well, I'll have a fresh start." I'm thinking, what'd you do back in the old place?

JOHN: You're exactly right.

MIKE: Do you have behaviors such that the people and environs are not sorry to see you go. So that you have to go somewhere to have a new start? I said, "Why don't you work on the things that made you believe that you had to move to a new place to have a fresh start." And you start to realize a life with many moves is a disjointed life—I don't know if you ever can get comfortable. I think it wears people down faster. I think it's more difficult on their life to keep doing that moving thing. It's alright to travel around, but as times change and it's easier to travel, it doesn't mean you have to keep on the move. Birds can point the way to thinking about, and perhaps changing, one's ways.

CHAPTER TWO

BIRD MESSENGERS, TOTEMS, LESSONS

The term "animism" has often been invoked in scholarly literature to describe a worldview—sometimes called a religion—associated with Indigenous hunting-gathering, farming, fishing, and herding societies across the globe.[1] The term is sometimes capitalized to place "Animism" in the same cultural category as world religions such as Judaism, Christianity, Islam, and Buddhism.[2] Thus understood, one of its defining features is often said to be the recognition of the natural, spiritually significant traits of nonhuman beings.

In the early days of anthropology, when scholarly writing on Indigenous cultures was pervaded with colonialist and white supremacist bias, animism was presented as an inferior worldview and the product of "less advanced" or "primitive" stages of human development.[3] More recently, the term began to circulate in scholarly and nonscholarly circles without condescension, and has been identified broadly with a concern to know

> how to behave appropriately towards persons, not all of whom are human. It refers to the widespread indigenous and increasingly popular "alternative" undertaking that humans share this world with a wide range of persons, only some of whom are human.[4]

Although neither Mike nor John uses the term "animism," I see strong resonances between their views and the animist worldview.

This chapter details Mike's and John's view of birds as non-human beings with their own personalities, and with important roles to play as messengers and lesson givers.

MIKE: There is a bird for every state or level of being a human is in. Some people really like chickadees, and they like the chickadee because the chickadee is the bravest little bird. Chickadees won't get chased away by the blue jays: they stay right at the birdfeeder. The other birds come in, but the chickadee just sits there and says, "I'm going to just keep eating." So, you can see how there's so many ways for a human to perceive a certain bird and whatever that bird has conveyed. How many people would walk around thinking that a chickadee is a brave bird? Not many. Yet the chickadee is such an incredible little bird and has so many messages to bring. The chickadee can be friendly and brave at the same time.

The kingbird is always the kingbird, the crow is always the crow, and they all have a message for all of us. Sometimes that message gets very clear, and it helps us move along. Someone asked me not too long ago about totems. If you have a connection to a certain animal or animals, that is your totem. Why would you limit yourself to just one? You have to use everything that is available.

When I make an offering, I ask every life-form, even the mosquito, to hear. If they can offer some bit of medicine or some ability to contribute to things getting better, you access every part of Nature. Not just eagles or hawks or wolves. And that is the part that I think a lot of people do not understand. You can use all of them as totems. All animals and birds contribute.

They all have a life force, and you can use whatever they can contribute. So you try to access the wind, the songs of birds, or whatever life force might be there. The life force can be directed or tapped into. Even a cockroach has a life force. All these bugs and birds have a purpose. They all need to be recognized for what they are doing.

Earlier I heard, and then saw, the belted kingfisher. I do not think I had ever seen him around so late in the season, what with eighteen inches of snow and freezing temperatures. I read that his territory is about six-tenths of a mile.

Was I part of the belted kingfisher's decision to stay late into the fall this year? I must be the only human somewhat regularly seeing him, as my house and surrounds form more than a mile-long property on the sides of "his" stream, which encompasses his entire territory and more. And there are no other houses close to his territory or mine. When he rattles around me it must be with me in mind, at least partially. So, he rattles for me, for Mink Hollow. He usually flies away from me when I exit my house or walk around the area.

DENNIS: I am fascinated by the belted kingfisher, the only kind of kingfisher we have here in the Eastern Woodlands region. Would we say that the kingfisher is my spirit bird or my spirit animal? How does one tell that a particular animal or bird is one's spirit bird or animal?

MIKE: There are a lot of factors, but I think one of them might be frequency of interaction.

JOHN: That's what I was going to say.

MIKE: The more you see it, the more you interact, the more it comes back to you. Cause you can't call them in. They're coming to you—that's an indicator. Because this is your bird.

DENNIS: OK. So now let's say there's a person who doesn't interact much with birds or animals and who doesn't go out in the woods. Do they have spirit animals or spirit birds, but they just don't know it? And are there birds or animals in the wild that are ready to be recognized by the human, but aren't?

MIKE: Yes. There are a lot of people who are not aware, because this kind of learning has never been offered to them. It's never been an area in which they would say that's why the earth is the way it is. That's why you're here, to learn from Nature. In a sense *THIS is school*. . . . And it's not a school just for learning, it's for interacting. And not only developing

BELTED KINGFISHER

Illustration by Julie Zickefoose

The belted kingfisher (*Megaceryle alcyon alcyon*) is the one found in the Northeast. There are several other species, usually brighter in color, throughout the Americas and elsewhere. A sharp-eyed fisher, it patrols ponds, rivers, streams, and lakes and perches patiently on branches or wires above the water, waiting to plunge for fish and crayfish. It often hovers midair with rapidly moving wings. "Alone, but self-satisfied," with a "sense of ownership," it "clatters up and down his beat as a policeman, going his rounds."*

Both males and females have a bushy crest, a large head, a bill longer than the head, and a shortish square tail. Slate-blue, with a white collar and underparts, it has small, speckled white spots on wings. The male, however, has a banded breast without the wide rufous brown found in females. Set in the grayish blue head feathers, both male and female have a small white spot just before the dark eye. This is one of the rare cases where female plumage is more colorful and complex than male.

Alert, easily disrupted from its perch by humans, it is often heard, but not seen, with a distinctive loud, mechanical rattling voice as it flies away. The female lays eggs in excavated tunnels in stream banks.

Fishermen and fisherwomen sometimes consider the kingfisher a thief, outfishing the humans, but the species' hyperawareness prevents it from being hunted.

In Western mythology and legend, a kingfisher is a symbol of peace and prosperity—the Latin species name being *alycon*. Indeed, an alternative name for the kingfisher is the halcyon. An old legend has it that kingfishers sit about in midwinter during the shortest days of the year when their brooding time is called the fourteen halcyon days, the seven days before and seven days after the shortest day of the year. The adjective halcyon still means calm, pleasant outings in Nature.

In some Native thinking, this bold diver reminds humans who avoid the new, or hesitate to engage in new opportunities, to dive headlong into novel activities.

Cherokee remark that the kingfisher was supposed to be a waterbird, but it lacked an appropriate bill until other-than-human Little People (akin to fairies or elves in European terms) gave the kingfisher a long bill in thanks for his having killed Blacksnake, who ate defenseless young flickers in a nest. Such stories— myths if you prefer—support the Native position that the invisible world of Spirit and Nature spirits are involved in the Creation and ongoing lives of birds. For Native peoples, the kingfisher, like all birds, like all life, is inherently, necessarily, a manifestation of the sacred.

* Blanchan (1917: 169).

wisdom, but relationship. If you want the better life it comes through understanding and relating to Nature and its interacting parts and relationships. These are concepts you can apply. The chickadee is your insurance paid up.

DENNIS: So each event, the appearance of this kingfisher or this junco that comes, or doesn't come, is part of an ongoing message center.

JOHN: It's constant.

DENNIS: If the junco comes and you're the only one here, is the junco for you?

JOHN: Yes. Basically, when the juncos come, and I'm here, they are literally telling me that I am carrying so much, too much, baggage because of what my mother went through. That's my biggest rock right now, that's my biggest thing, to try to accept and understand, without any sort of anger, without any sort of hatred toward anybody.

DENNIS: What's interesting is, if you see these events of Nature having messages for each one of us, then we're being more open to ourselves and exposing ourselves more. I think many people spend a lot of thought and time covering up their shit.

MIKE: That's what makes them sick.

DENNIS: I mean, here we are, I don't even know you so well, but we're pouring our hearts out. We respect one another. But this doesn't happen very often. I'm kind of an open person, I think, as you are. Part of the reason you're here, and we can talk, is because we're these kinds of people.

JOHN: But the system teaches us not to open up.

MIKE: There are situations when birds have to mask their stress. Cause if they show vulnerability they're done. They might deceive or hide for survival, not for any other reason. They're not trying to hide the fact that they're vulnerable because they're afraid of what other people might then say, "Oh, you are just a robin, you don't have big talons and a sharp beak. Well, you must be foolish." But you can watch how they protect the young. When the fledglings come out, one of the parents will hold their wing funny and hop sideways and act like they're injured. When I started to watch the birds at Hawk Creek rehabilitation center, the birds will cover up injuries to make them look like they are fine, because if they show vulnerability, they will probably end up as food.

DENNIS: Does everybody have a specific bird totem, but just doesn't know what it is? Who's choosing whom?

MIKE: I think the choice is made on both sides, by the bird and by the person. If the person does not have an interest, then the birds might not come around very much. Marsha likes the colorful birds. So cardinals come to her feeder, and Baltimore orioles and goldfinches, and other finches. Birds with bright colors come because she has an interest in them. And she seems to have that rapport when we're driving down the road. I'll see the crows, the hawks, the vultures. I think they show themselves to interested people. When you have an interest in seeing a particular kind of bird, they seem to cross your path. So I think there is, in a sense, a choice on both sides. The birds that you like to see will show themselves. This should be an indicator for people that these are the totems that we should pay attention to.

For example, the catbirds. You can hear them, but you hardly ever see them. You always hear the call. And then you're looking to see what kind of bird s/he is, and they often keep themselves hidden.

DENNIS: It's funny. I try to sneak around to see the catbirds, but they always go around the other side of the tree or just fly up to the branch behind the leaves that you can't see. It's as if they know exactly what they're doing, hiding from you.

MIKE: They do. That's what they are about. They are about not being in the human's line of sight. That's what they do. Cat-birds are really good at that. You have to really look for them.

DENNIS: The catbird is often hiding, almost like a cat and mouse, or human–bird game. Right, it's a bird–human game.

GRAY CATBIRD

Illustration by Julie Zickefoose

The gray catbird (*Dumetella carolinensis*) gets its genus name from the Latin *Dumetum*, meaning shrub or bramble, as this bird often inhabits thickets or brushy areas, usually five or six feet off the ground, especially near water. Elegant and lithe, it often flicks its wings or tail, and usually flies short distances, hopping from one branch or tree to another, sometimes on the ground, sometimes catching insects midair. Catbirds are omnivorous but prefer fruit and insects. Their plumage is almost entirely a slate gray, but with a marked, small black cap, black bill, black eyes, and black legs, and a few chestnut-colored undertail feathers. It has a fairly long, rounded tail, which it often spreads.

Although it often hides from direct view, as it quickly flies behind tree trunks and branches, the catbird appears to like being relatively near humans, as if playing interspecies hide and seek. This vivacious bird usually lets nearby human observers know it is around, voicing its namesake mournful, mewing cat call as if complaining or protesting, or, as some think, specifically to annoy humans. (A human squeaking a call or kissing her palm may well summon the bird from afar.) The catbird has a vast array of melodious songs, mimics other birds, and offers a medley of unrepeated phrases. The catbird has seven pairs of intrinsic syringeal muscles, which account for it being a diversified songster. As well, it can sing with two voices at the same time due to the two-sided structure of its syrinx. (One birdsong authority listened to one catbird and heard 74 different phrases.*)

Andrews† suggests that the catbird is a "busybody." "Its presence should caution you to be extra careful about what you say and to whom. . . . Its presence can hint at others being overly inquisitive about your own affairs or that you are being so about others. The appearance of the catbird indicates that some new form of communication is going to be learned. And the presence of a catbird as a totem indicates you will be encountering a wider range of people . . . who will teach you lessons in your ability to communicate."

* Collins Jr. and Boyajian (1965: 121). † Andrews (1996: 125).

MIKE: I think it is kind of a game. The hermit thrush is another bird that is shy. . . . And there is a story about that. . . . It has been said that in the beginning when birds were getting their assignments [from the Creator] the birds rallied and vied for positions because it was kind of a competition. But the Creator put a very beautiful song up in a cloud. Whichever bird could get to that cloud would get that song. For days, all the birds were out practicing their flights, getting their flights ready, seeing how high they could go, because the cloud was pretty high up. So when the Creator said "today's the day" when that song is going to be awarded, the eagle knew that he could easily fly up to that cloud. All the other birds were watching in preparation, getting ready to take off. But the hermit thrush really wanted that song and so the thrush flew in under the eagle's wing and hitched a ride. The eagle didn't know the thrush was there. All the other birds are flying hard and fast, trying to use the currents and thermals. But the hermit thrush knew their wings couldn't take them up to that cloud. All the birds except the eagle started to fade off. Just as the eagle was making a final approach to get to the cloud the resting hermit thrush flew out from under the eagle's wings, flew ahead of the eagle and got into the cloud. The hermit thrush got that song. . . . And they say the eagle screeched because the eagle wanted that song, but the hermit thrush got the song. That's why the eagle screeches.

DENNIS: Wow!

MIKE: And that's why the hermit thrush has this song. But because he felt ashamed of what he did to get that song—that he didn't actually earn it through his own flight—the hermit thrush stays hidden and won't show himself. But he has a beautiful song.

DENNIS: Some birds sometimes look at us directly. Some look at us as if they're flabbergasted at our mentality, like they're psychologists looking at how crazy we are.

[Laughter]

Others stay out of sight. . . . I am often surprised when I am trying to get a line of sight on some birds, and I want to see them closer and talk to them, how, even if I'm quite a distance away—like looking at a woodpecker a hundred feet away—the bird, from my view, goes around the back of the tree.

The house wren landed on the front stairs' iron handrail, as it made distinctive chirps. Only the screen door separated us. I could see him turn toward me, look quickly at me, and then fly away. It was as if he came to see what I was doing.

MIKE: Squirrels do the same thing. They'll try to hide from your line of sight. I think a lot of it is a bit of a carryover because of what hunters do. But certain times they get more comfortable and won't stay hidden or try to hide.

The birds are one of the first indicators of Nature's acceptance of you. Everything out there is using all the other animals. So if the deer all of a sudden see birds fly away, they start looking around, saying to themselves, there's a dog or something else around here that we better keep an eye on. It is interesting to use the woods or the animals to see if you are accepted by them or can be a part of what they're doing.

Some people think it can happen quickly, to be accepted by the animal. That's a nice thought, but you need to spend time out there to learn how to behave in their place. You are in their home now. That's their place.

DENNIS: When they pick up on our energies, what is it? The human's attitude, nervousness, aggression, fear? Or what?

MIKE: It could be all of that because humans hang onto a whole series of energies, emotions, and states. Nature doesn't have these kinds of emotional states. Animals do have a few responses if they're on higher alert, but it's really not feared the way a human defines or understands fear. Humans can create their own fear in their minds without an immediate threat. When a fawn is born and the doe is licking the fawn, that's not affection. I mean we may want to think that it is, but the licking is a natural response to help the fawn, not only to bond, but also to help stimulate it to get active. You can't be vulnerable for too long out there or you're dinner.

Outside the window a mourning dove flies by. Then another one.

DENNIS: Sometimes I see a couple of Canadian geese by the pond next to my house. They usually squawk when I'm nearby. Are they squawking to each other or are they trying to communicate to me? You are suggesting that that squawking is not an emotion.

MIKE: It's a response to the surroundings. Whether warning each other that there's somebody down here, or something else. When the geese are in their flock, they are very good at communicating. They do a really good job of looking out for each other. They really are a collective group. It's interesting even to watch starlings fly in a large group. They do some bizarre and beautiful movements.

Later, Mike continues to talk about the collective behavior of some birds:

MIKE: Seagulls are stubborn birds. They are not like the crows who keep the families together. Seagulls chase hawks or owls or anything they feel is a threat. I guess they figure that they eat anything. But crows definitely have a more collective order to how they function. Crows are much more organized, and I guess you might say of a higher intelligence, or higher function, than seagulls.

But seagull behavior is changing. In the last few years, they form Vs about an hour before sunset, heading back toward the waters. They are becoming higher functioning, proven with geese that form Vs in their flight patterns. Ducks will do it too. They form a V because the head, the leader, breaks the initial wind force. It's like turbulence behind a jet plane — once they break the air current, the rest of them fly within and they sense what the angle is just right for their wings. That is why the lead guy switches often. They move back because leading tires them faster, and they move back into the V. Another one takes the lead.

There is so much stuff going on in bird and animal behavior. I do not like that science typically will say that it is instinct. You can call it instinct, but I see it as a higher intelligence left to a higher function.

The structure of a bird is incredible, and flight is almost a miracle. Birds' bones are so light. The air pockets in the bird bones are so large and yet the bones are strong. I mean look at falcons. They have to be strong because they hit their prey at fast speeds.

Humans do not compare with the birds. The humans have weapons and shields and protective things. Birds do not need them. And pound for pound, humans do not stand a chance. But humans consider themselves superior.

DENNIS: Oh, one of the most fantastic things is when you get a big flock of starlings, or any set of birds in a flock. How they follow each other in the sky so very closely and accurately without touching one another! It's a moving sculpture, and the individual birds stay the same distance away from each other. And in a fraction of a moment, the whole moving sculpture can change direction, up or down, and they're in perfect unison. It's incredible. That's one of the greatest scenes.

Those flock flying scenes are called murmurations, and scientists do not know how to explain their behavior and expertise.

MIKE: But that shows you that the life out there has a collective element to it. They seem to be unified. You go "how the heck can they do that." I mean being in flight like that and then they just change directions so quickly, or all go into a spiral.

DENNIS: The starlings and other birds are fantastically aware of each other's space.

MIKE: It's an amazing maneuver to see that happen. It's fascinating to watch those flocks. And you know I'm just amazed. Is there a real purpose to it? Is there some other function that they're working on? There's got to be. It's not to entertain themselves. I think it's something that attunes their own abilities to the group's frequency.

DENNIS: And when geese are in flocks in flight, sometimes they change direction and then they have to realign themselves into either the same V or maybe two Vs and sometimes you can see it happen right overhead. It's like they are all engineers realigning themselves. And they stay the same very short distance from one another while they negotiate the forces. They're readjusting all the time.

MIKE: It's interesting that they understand aerodynamics. They really do. They have to. To migrate, the birds' lives depend on getting to a warmer climate or area, and they might fly a whole day, or several days, without stopping. All of the behaviors are a part of that. How did the first generations establish and know how to pattern the things that they do? It's really so incredible that each species has their own behavior.

MIKE: These things—flocking, varying bird behavior, etc. are some of the greatest teachings. Some of the greatest education is observing in the wild.

These past few weeks, since the days have gotten much shorter, I have felt out of sorts, out of tune. I realized that it was greatly a function of not spending much time outdoors. I began to resign myself to feeling less embedded on a daily basis with my bird friends and wildflowers and green things all around here.

But I just got back from a walk in the snowy December environs and had three or four bird experiences, which showed me that even when there are only a few species around I can get avian satisfaction and jollies.

As I was walking down the dirt road just after the bridge over the creek, I first heard the rattle, then saw my first totem bird, the belted kingfisher. It must have been more than coincidence that he flew and rattled nearby. Behind me at the beginning of my walk, just after I passed the creek. I thanked him and the Bird World for letting me know he was around and mindful of my presence. But I wondered why he was still here in this cold, wintry weather, why he had not migrated. I do not think in twenty-five years I have ever seen or heard him this deep into the winter season. What led him to stay?

Then later I saw and heard a small flock of slate-colored juncos, with one still on the ground on the snow on my left, as I approached. Most of them had already crossed the road and were in bushes and small trees on the right side of the road. That was almost exactly where I had decided to turn around and head back to the house.

Then, not far from home, in the last patch of woods along the road before the fields by the house I heard the distinctive call of the white-breasted nuthatch. I didn't see him, but his voice was loud—he must have been saying hello to me.

I am back.

DENNIS: This year the belted kingfisher that lives along the stream next to my house is staying late into the fall.

JOHN: Yes, but the water is still open, not frozen.

MIKE: The waters are still open. He'll or she'll go when things get really cold. They won't stick around when it gets cold, they'll just wait for it to get cold.

JOHN: That's one of the things with the water temperature being so high: the belted kingfisher is staying longer than what it should.

MIKE: That does pose a problem in the sense that when they migrate, they're not going to find open water in other places that did not have the extended warm weather, because elsewhere many waters are already frozen over. They have to travel quite a distance to get to open water again. . . . I noticed all the hawks are still here; the red-tails are staying longer.

JOHN: A lot of hawks.

DENNIS: And ordinarily by now I would see more activity among blue jays. I don't think I've seen any lately.

MIKE: It's just part of the whole way that system works. It depends on the weather to initiate those changes, and that's how they respond. One of the things that I've noticed is when the robins start leaving, you hear and see a lot more activity from the blue jays. The blue jays are not what most people think or see. They think that the blue jays are aggressors and a dominating bird, but when the robins get here the blue jays become very quiet.

DENNIS: In the spring you don't see blue jays much.

MIKE: They are almost nonexistent then. The robins are the ones that become the dominant bird. But people say blue jays do this and that, and we see them at the feeders, and they kick the seed all over. The blue jay actually gets less vocal in the spring, and the robins seem to take over that time and become dominant in that space. But the blue jay is not forced to take a back seat, he just does that. But why does he come out in the fall? Because part of his job is to warn the other animals that the humans are going out there to hunt them, and that's been happening for so many generations and now is an automatic response.

DENNIS: Let's say, 300 years ago, before the white man took his shotguns into the woods, you're suggesting that the blue jay was less vocal?

MIKE: He may have been because the Native people were more respectful in the hunt and in all the environment.

JOHN: Yes, everything was working much more for harmony with each other. Whether it was man working, or blue jays working with man, instead of scaring deer and other animals.

MIKE: Now the birds and other animals are more on a defensive position. The mainstream, or if you want to narrow it and say the white mentality, looks at hunting as a sport. I'm going, "you're going to take a life and you call that a sport, I thought football was a sport." It becomes so violating to life to call hunting a sport that it is quite understandable why these other animals take their roles more seriously now and have to warn others.

DENNIS: That's interesting. The birds are actually more serious than they used to be because they're more on the defensive. They used to be freer. Freer in the sense that they were less worried about their own survival and less having to do a job of warning the other birds and animals that their lives are in danger.

MIKE: I guess humans want to personify; they want to humanize everything out there. To most humans the crows are no good, the seagulls are no good, the owls, well, we're not sure, but the hawks and eagles are "good." That's how they want to see things. They're not looking at it from the perspective of that life. In the last few hundred years, they never have because the dominating humans and leaders have been saying that they are in charge.... Nature has been so transformed. That's why we're in the state we are now, because they have detached from Nature. These bird, animal, and climate happenings are cosmic events. Humans didn't invent them. Some human didn't sit down and write this out in a script.

DENNIS: A blue jay, I just heard a blue jay [corroborating our words].... So, John, as Mike was saying, it seems that the blue jays, from a human perspective, are aggressive birds. They hassle other birds and tend to kick birdfeed out from the birdfeeders humans put up. So, if it's not aggression, why is the blue jays' behavior more like that than some other birds? What is it that makes the blue jay have that role, versus, say, the chickadee who seems to be everybody's friend?

JOHN: I don't view the blue jay as any sort of nuisance.

MIKE: Lately because of all the white-tailed deer in the area and humans infringing on the deer's territory, taking more of their deer habitat, now there's a lot more deer/car collisions. The Department of Environmental Conservation, the DEC, is having off-duty cops bait and shoot the deer. I went up there to burn tobacco and guess what came in? All the blue jays.

JOHN: Sure.

MIKE: And I said you know what the role of the blue jay is? To warn the rest of the life out there that there are things being plotted against them. We need the blue jay even within the human group. We need blue jays that can recognize what's being done that's not right and go warn the people. That's what the blue jays do. . . . Well, they're loud. They can be viewed from the human realm as obnoxious—they like to hear their own voices all the time. But the part that I see of them is they warn all the other life out there and warn that somebody out there has a gun. When they see somebody with a gun out there on their shoulder, blue jays go out and start telling everybody, "Hey take cover, they're coming to get you!"

DENNIS: Do the blue jays also bring messages to humans?

JOHN: Oh yes.

DENNIS: Are they of a particular kind because of the nature of that bird?

JOHN: I think the message is on the lines of "Look at yourself."

DENNIS: Like you were saying, the message to the human is "You are arrogant." The bird is trying to say, "Hey man, you think you're a hot shot, but look at me. You are not much compared to me and my connections."

JOHN: People often complain because the blue jay will knock the seed from a birdfeeder onto the ground, but your juncos eat off the ground. They're actually helping the juncos and other ground feeders.

MIKE: When the human puts explanations of bird behavior into humanized versions, it's usually wrong.

JOHN: Isn't that right.

[Laughter]

It's true.

DENNIS: What about woodpeckers? Blue jays tend to be aggressive. They have bold coloring that gives us teachings that you guys are explaining. What are the specialties of woodpeckers? There's the hairy and the downy, the pileated, and the red-headed woodpeckers, among others. The flicker is the only one that feeds on the ground and is very much multicolored.

MIKE: I've noticed that the woodpeckers have become much more prevalent. There are a lot more of them now. . . . When I was a boy it was a treat to see a woodpecker come out to the feeder. Fifty years ago, you never even heard of a pileated woodpecker. They were not there.

JOHN: The red-bellied woodpecker was also never around.

MIKE: And even with the overuse of DDT, and other things, many birds have found a way to overcome that to survive. These are all of the components that teach us. When we make a reference to the animals talking, we mean that they are talking not only in their own language to other birds and animals, but they're also talking to us, telling us about their health conditions based on what's happening in the environment they share with us. Those are key pieces of information.

That's why what John is doing here is so important, to bring this to light to even the DEC. Disappearing birds and animals, and other changes came because of the changes humans have made, including the DEC, and the things introduced into this environment. Thinking that you can dig a hole and dump whatever it is that you think you can hide, to get rid of. But it's harmful to your life or others' life. Or you think you can sail out over the water and dump trash and chemicals way out there and it's OK to do that because there are no intakes over there. You have a high IQ, but you are dumping things over here that you've already determined to be harmful to life.

DENNIS: So, are you saying that the expansion of species of woodpeckers is a result of human tampering of the environment?

JOHN: One of the things that I believe is that the big increase in the population of woodpeckers has to do with the amount of tree beetles. And there's the emerald ash borer, killing thousands and thousands of trees. . . . This is part of the bird prophecy, that this northern part is really going to get hammered with insects, tree insects, that are going to kill the trees. These woodpeckers are part of the army that's going to keep it back to a balance point. . . . Also the woodpecker to me stands for the drumming of the Earth Mother, of her heart. Every time the woodpecker knocks on a tree. It's unfortunate most people cannot recognize that, because if they did, maybe for two minutes, they might think about the Earth Mother.

DENNIS: The drumming brings attention to the Earth Mother. And it's one and the same thing—honoring Earth and finding food.

JOHN: That's the way that I feel. Like Mike said, all of a sudden, we have all these large woodpecker populations, and they are coming to feeders, which they never used to. I remember having a feeder over at my mom's house when I was young, and you never saw them.

MIKE: Earlier, I think many trees had been cut down and that's what devastated woodpeckers. Their food source and habitat were destroyed.

DENNIS: The trees are starting to regenerate more. There are older trees now.

MIKE: That's good. They need to leave some trees to get large, to get to that stage where the woodpeckers can use them. And look at the relationship of woodpeckers that make holes that squirrels can use, and raccoons can use. Everybody benefits from what they do. You see again how it's a harmonious system. . . . Woodpeckers are the construction workers that make homes in the trees for the other animals. Connection to, and harmony with, the other elements and birds and animals *is* their system, and their sacred lesson to us. So you start to see that the connectedness is sacred—that's the whole key. But you can't force sacred on anybody.

Doctors and most people don't put any credence in what the birds are saying, or what they are singing. Yet that's some of the biggest medicine.

If your doctor wrote you a prescription to listen to bird songs for a set amount of time it would give you relief from anxiety, depression, and other ailments.

DENNIS: Chickadee therapy!

MIKE: In conjunction with whatever you are thinking about at the moment when you look and see the hawk way up there, it could refer to a relationship that's really distant. Or it could be other things, but what's important is what I get from the hawk.

JOHN: There is no cut-and-dry rule or linkage. . . . Sometimes people will bring animals over and ask me, "What's it mean?" Well, "What does it mean to you? Let me hear your thing first."

DENNIS: So the hawk flying high is one example. Say I'm thinking about something, bothered about something, or planning something. It's there at that height in part for the hawk's own life, but also to act like a tarot card for the humans who see and connect with the bird. I can use it to learn about myself or to direct my own thoughts or behavior in some fashion.

MIKE: Yes. Well, I know that my distance vision is really good. When I look up, I say to myself, "Why did I pick that hawk out?"

DENNIS: That's also where sometimes I go in my mind. I try to figure out what's happening, but maybe I'm trying to figure it out too much.

MIKE: Sometimes, in some instances, it can be a problem. When we are overthinking. We overthink too many situations. Sometimes we muddy the water or cloud things up by becoming too human. Using too much from this man-made system, not from the natural world.

DENNIS: Even if I over-read the birds, I don't feel like I'm injured. But when I'm with a human being, if I read into their behavior, or my behavior, it does become sort of self-defeating. I can't imagine that my reading into the hawk's flight could injure me even if I'm being overly analytical. But if I think about your motivations for why you are talking to me, or something like that, it could screw my head up and interfere with my connection to that person. . . . The great thing about Nature and the birds is that they provide so much richness, so much richness right here, right now, right as you open your eyes. . . . I know a lot of city people who easily get bored with the trees or the birds. But there is so much right here, right this minute.

MIKE: This understanding is just what most people have drifted away from. We've stigmatized the Indigenous type of belief system. Yet the Indigenous belief system followed and recognized the structure of the Creation. Some believers base their doctrines around a person who has come in at a certain time. Religions like Christianity do point to spiritual existence beyond material life, and such a view is part of biblical teachings. But some of its individuals seem not to follow the teachings, but rather follow television personalities and evangelists who have taken much of the spirituality out of the religious group, sometimes defined by who they hate.

DENNIS: Like a celebrity culture we live in, even for religion.

MIKE: Yes, even in modern times among Native Americans. There was Sitting Bull, Geronimo, Chief Seattle, Handsome Lake, etc. There are a lot of renowned Native people, but they didn't come here to be famous. They came here to give a message. And the message was based on what they were experiencing during the time they were here.

If we're all from the same Creator how is that going to make the Creator happy if one group is destroying another part of this Creator's Creation? You might say that the Creator loves his Creation, which is true, and that is also I believe part of Judeo-Christian doctrine, but some human groups think that they're better, or that their way is the only way. So you start to see conflict. It's because few focus on the Creation or on Nature, saying, "Wow! All our answers and all our models are right here."

It baffles my mind. "Come on, you really don't make the connection? You don't see that every solution is right here?" Yet it's built in. But people want to humanize the animals, but they're not human. First of all, start respecting them for what they are—start watching crows and start watching owls and hawks and birds of prey. See how they interact. But there's a bunch of humans who will think that if they were that hawk, they would shred a few of those irksome crows.

Nature will show you how harmony works. You'll never see one group gang up and say we're going over the hill to take on that other group to get rid of them. If they don't like each other, eventually what happens is exemplified by the crows and the hawks. The hawk moves away finally, and the crows finally leave them alone. The crows say to themselves, "We just don't want the hawk near us." A crow will not even eat or scavenge on a bird of prey. It is just part of the structure, the nature, of things.

DENNIS: That's interesting, so not only will the hawk not act aggressively against the crow when they're alive but even when they're dead they won't have a relationship.

MIKE: I asked John if he ever saw a crow scavenge or eat a bird of prey? Eat a hawk? I've seen lots of hawks on the side of the road, but the crows won't go near them. In fact, what triggered that was that we picked a hawk up just around Christmas time and I brought the hawk over here. About two weeks later right in that same spot I saw a crow fly over that place and that's when the crow triggered that question: I've never seen a crow scavenge a dead hawk, even if crow scavenges other animals.

One time the three of us were talking about how individual birds could have different personalities. Outside John's kitchen window, we heard and saw birds comment on what we were saying—a blue jay had been monitoring our conversation.

MIKE: Yeah, just now. Look, a blue jay flew in, and threw that message in there. It's almost like the confirmation that says, "Yes, we're all a little different." The blue jays have general characteristics, but if every blue jay behaved as only the general characteristics, then you wouldn't have diversity, or have some that become exceptional or specialized. Hunters do not like blue jays because when they walk out into the woods they start talking. Does every blue jay do it? No. Otherwise that's all you would hear out there. So it's only a select few. The catbird also calls a lot around humans. When John came over to help stack wood, a couple weeks ago, we went for a walk in the woods over by this guy's house and a catbird followed us all the way around as we walked the property line.

DENNIS: Complaining that you were there?

MIKE: No, it's just an observation. But the catbird will let you know he's there. So, you hear him, and so we're down by the pond and after a few minutes the catbird shows up. You can't see him, but you hear him making his call. John noticed with himself, and I have noticed about myself, when I go into the woods a catbird will always come close by. Especially if I'm burning tobacco. Like the blue jay, there are a few catbirds that are exceptions within their group. Those are ones that come over and make a connection. You start to see that—it's not like you get twenty catbirds—there's only one that usually comes around and you only hear that one.

JOHN: Even when we went to the pond to get some cattail roots Mike needed to help heal someone, as he was getting the cattail root, sure enough, there's the catbird.

MIKE: We have to use all of these birds, not just as some people pick favorites like "I want to be the hummingbird" without much concern for all the other birds and elements. We have to take what the eagle has too. That's when you start seeing each bird species has their own specific duty. And they all work together when we start watching how they all interact. Each one has their role, a specific calling, what they need to do in life. So, you start to really see it is a harmonious system out there.

And even within the species each individual is different. . . . The pine tree is the same species as that other pine tree over there, but when you look at the detail, they are not the same.

They're growing in their own space at different rates and doing different things, even though many humans don't see that. Often, they will look at birds this way and say, "Well, ah, there's a nuthatch, he's just the same as all the other nut-hatches out there." Well, like with humans, they might all generally look similar, but of course they're not all the same. A specific bird may have a specific role in a person's life.

DENNIS: John, do you find individual differences and exceptional qualities in specific turtles also?

JOHN: Yes, you find it in all the species.

DENNIS: What would be an exceptional quality or behavior of a turtle?

JOHN: Well, when they do the dance. Wood turtles will actually do a dance. Not all of them do it. To me, that is something very special. We have one turtle, and you may say that it's a deformity, but the turtle had an extra toe, which to me is really important. . . . There are so many different things, you know, and it is really hard to pinpoint, as just one of the species suddenly starts to talk to you. . . . When we get birds, hawks, owls, and living things, each one that comes to us is extremely special. They come for special purposes. It is fantastic when they give you that message. People will look at you like you're nuts.

And the same thing with the deer racks. When people bring their deer here to be mounted and you start reading to the people the medicine of the deer rack, and the people say, "Wow!", everything fits. It's like, "Yes, that's because that deer picked you for whatever purpose the deer had, or to teach you that your female side is not a good side. How you treat females is poor because that side of the rack is all messed up." Different things like that. It's just amazing.

DENNIS: Is there an explanation? Can you figure out how it happens that the deer with the messed up right side of the rack actually picks that particular person? How does the deer pick that person?

JOHN: After walking this path for this amount of time, I'm realizing that everybody who I've come in contact with has the choice to walk a good way, or to walk the way of the world. Everybody has that opportunity. But I think a lot of people have to be shown it, they have to really clean it out and be told "Here, look at it." The right side of the rack is fine, OK. Then you go to the other side, the female side. I had one rack that was really deformed, and the guy asked me about it and I said, "How do you treat the women in your life?" He had his girlfriend right there and she turned around and walked outside. She was scared. He was abusing her, but she didn't want to see it. And I said, "Well, there you go." I said, "That's why this deer picked you. To try to get you to understand that just because she's female she's still a human and you don't mistreat her. Just because you're a man doesn't make you superior whatsoever."

This conversation reminded me of my fascination with specific visitor bird experiences:

DENNIS: One of my favorite moments that I keep playing in my mind from many years ago is when a hummingbird came to my porch. There's a big glass window, and I know he looked me right in my eye through the window. He turned his head like this, turned his head like that, and I was looking right at him. I saw into the eyes of the hummingbird, and he into mine. I still fantasize on that. The experience was so powerful. It felt like I was looking into the eye of God, or the Creator.

MIKE: Yes. The word powerful applies, but most people's definition of powerful is, well . . .

DENNIS: 450 horsepower.

MIKE: Yes, that's it. They don't look at "powerful" as being a life force that doesn't need to be physical. Sometimes life forces are nonphysical, and they are powerful. And you start to see, how does the wind categorize itself? I think the wind is both.

I think it's a powerful force even though you can't literally see it, but we know it can take trees down and we know it can move almost anything. And you go, but it's not visible to the eye. The only thing that makes it visible is when the trees move. I think the wind is a mystical, magical powerful force. An invisible force, it almost transcends physical beings, but can really interact on the physical level.

DENNIS: When, for example, the junco comes, he is active medicine. Not only can we read the bird but the bird himself has active medicine to give, if one is receptive to it.

MIKE: Each species has their own message and medicine.

DENNIS: But let's say I don't really know the medicine. I have my own things that I read in birds, and of course I don't know the birds' medicines like you guys. If I don't know, does it still affect me?

JOHN: I believe it does.

DENNIS: Am I or somebody else learning from the junco even if I don't have the idea that I can learn from the junco?

MIKE: To a degree you are. But it gets enhanced when you become cognizant of it and make the connection to it. That's why knowing what plants do, what they are supposed to do, enables me to make full use of them. I talk to the sage. I used to watch Mad Bear when he would make a pot of sassafras tea and talk to it in Tuscarora, and I asked, "What are you saying?" He said, "This sassafras knows what it should do and I'm just reminding it." So, we talk to it, and tell it its job, this is what sassafras is for. "This is why you're here, be grateful for that because you are medicine." He really explained some of these things. I go, Wow! That doesn't just apply to sassafras, that applies to everything out there. So even when you see crows or vultures, you can talk to them.

JOHN: Well, this is the part where they don't understand our way. They don't understand that type of medicine. All these helpers come here for a positive reason. That junco we had wasn't just for me, it also was for my wife, and other people who come in. We use these helpers, and everything that is sent to us. We use it to show other people. "Look we have a junco right here—that's telling us *ba-da-ba-da-ba*, something you need to look at." The object of this whole life thing is to get rid of the nonsense that has been driven into us and to become more and more connected. I think it's just that simple. . . . There is responsibility when you take on the [sweat] lodge, the responsibility of helping Nature and people. People who need help are going to come. They can't get that kind of help through the health insurance and hospital system.

You know people come here because they like the idea of what's going on with nature, versus their own life. They feel like they connect, instead of being pushed aside by the system. Everybody used to benefit from the medicine that was in this house whether it was a mounted bird or animal or if he or she were alive. Those DEC officials who took away the stuffed and live birds and animals didn't just hurt me, they hurt a lot of people.[5]

DENNIS: Did you have a sense that before these DEC guys came and took the junco away that the junco had different behavior?
JOHN: No, but the goldfinch that Mike brought over actually died. Goldfinches to us, because of them being so bright and yellow, stand for a brand-new beginning. And reminds us of the sun, and the direction of the East. The day before the DEC came is when the goldfinch died, and that bird had been doing great.

GENERAL PHILOSOPHY AND PRACTICE

DENNIS: Spring is a fantastic period. So quickly so many birds and plants grab your attention. It's hard not to marvel in it. This time of year, so many things are happening at once. I swear I can feel the energy coming out of the grass as you just look there. You just feel it like a force field of energy.

JOHN: If more people would get close to people who spend time in Nature, or spend time in Nature themselves, more people would really start to evaluate themselves and say wait a minute, I have to change this and this and this. Will people be able to have birds and animals come to them, I do not know. But Nature helps to bring people closer to themselves and others.

DENNIS: I wonder if the people who get freaked out when they go into the woods or see a spider, or whatever, if what is really happening is that they are disappointed at their lack of connection. That they do not really know how to relate and be comfortable. So rather than try to learn, like some of us, to be closer to Nature, they move back because they are either afraid to be closer or they just realize that they cannot get close. So they retreat.

MIKE: Well, there are a lot of reasons why people do not attempt— often because they don't want to fail. . . . You need to know that you have to fail before it happens. You have to fail. You just cannot go walking out there and it is going to happen. You have to be prepared to say that it is probably not going to work for a while. But if you do not give up you see that it does happen. There is no accelerated course. The failures are part of the process. . . . The Mohawk language is especially attuned to experience in nature. Mohawk is like watching a big screen at a drive-in theater. The Mohawk language is like looking out the window and seeing all the languages of the trees, the language of the grass, the insects. The video screen is OK for a few things, but how do you relate to the wind blowing through the trees and the temperature with the sun, all these combinations of elements?

I've counseled people on prescription drugs with anxiety and depression to go out into Nature to cure their dependence on drugs. It works. . . . It's what I do. Try to bridge the everyday with the abstract, which the Native people don't see as abstract. They don't see the wind as being gusty, or shifting directions, as an abstraction. We look at the characteristics of this tree versus the tree across the walkway. Or these smaller ones over here, and what they do, how they move in the wind.

And not just the air. If we walked out there and we had an unsettled attitude, our mindset could change with the wind. But maybe we needed a calmer day, maybe we needed no wind. That's the stuff that we have to pay attention to. I was mentioning to Dennis earlier that humans are probably the worst piece of creation on the planet because they're unstable, they're fearful. They don't cope well. If there are no monsters, they'll create them. They have insecurities that run deep and turn into life patterns and psychosis. What's probably the biggest industry going right now? Therapy. Nature is actually out there for our benefit. All of it, when we see it that way. Then Nature becomes the Medicine, the thing that will actually help us. If we start watching animal behavior, not animals confined by humans, but animals that are out there to live in the environment that they're supposed to be in, we will learn every teaching or everything that's necessary for us to use for relationship. And when you hear the word relationship, most people think of men and women or other humans or maybe a dog or a cat. In our version, we perceive relationship as how you respond to every element in this creation.

So every element becomes important. And even as insignificant as you might think it is, and may not recognize, Nature provides some element for you, either on the subconscious level, or actively to recognize. When you start seeing or hearing, even to watch the way some of the leaves respond to the wind, these are all parts of what we should be looking at, taking in. If therapists started using Nature as part of their treatment, it would put them out of business because the people would become adjusted in using nature as the medicine and as the therapy. Whether it's animals or plants or insects, all of them will demonstrate to us that these life-forms are to be observed by us. They want us to see what they're doing, and they want us to observe and pay attention and say wow! There's something I haven't seen before.

DENNIS: Does the wind have a spirit or soul of its own?

MIKE: It does. And the wind is a very useful thing for many parts of our practice. One of the things that we know that works is the wind or breeze as a distributor of sacred tobacco smoke. When Native people say we need to burn some tobacco, it is not the same as another person with a pack of cigarettes saying I'm going for a smoke. When we offer tobacco, we are putting it out there, asking for these life forces. The wind is one of the life forces.

DENNIS: So the smoke goes where it needs to go?

MIKE: Yes—that's part of the purpose of the wind, which carries the smoke where it needs to go, whatever the message. Do I know where that smoke has to go? No. Does the wind? Absolutely. That's why we need, and we use, the wind to carry that message, to carry that smoke to where it has to go. Does that message have to go to an individual or is it going to an entity or another life force that's out there? Or does it go to all of it? I get feedback later. People will ask something like, "Were you burning tobacco earlier today? Were you sending a message out or were you asking for something?" I say, "Yes, you're really perceptive." It's not usually the youth, it's older people who allow elements, and thoughts, to come in. Or something changes. That change can be just the wind by itself. The wind, like the birds, has flight, and helps bring messages or lessons.

DENNIS: The birds have different niches in the same tree, right? One goes up one goes down, like the nuthatch, ones at the top of the tree, like the cedar waxwings, ones at the bottom of the tree, like some sparrows. And they get along on the same tree, literally the same wood. Humans don't seem to do that.

MIKE: Well, because when groups hold the concepts of divide and conquer, and preside over others, they think it makes them superior. This mirrors their beliefs that some animals, or some birds, are much more important or powerful or dominant over others in a system of moral ranking. And in their alleged superiority over other humans and animals, they are operating at a much lower level of functioning than one could be. A level that is about ego or greed or power, rather than harmony, integration, honoring the spirit and practice of Creation.

And Nature has no violations or untruths to hide. None of that life has to deal with this problem of right and wrong, true or false. This is not to be confused with protecting or shielding sacred elements, but now we are talking about offenses against nature which humans often try to hide. . . . What do blue jays want to hide? None of that bird life has to deal with this issue of hiding improper behavior. That's when you start asking about Nature, more than the nature of humans, asking if Nature provides great things that are gifts to the humans? Absolutely. But the sooner you get to that connection of Nature, not to hide, but to align with Its natural design, then you have to take initiatives to try to stay on the side of "I don't have anything to hide." Here is a great lesson from Nature.

So, every little piece of what is out there not only has a purpose, but It does what It does at the right time. The birds hold a purpose, and they are all listening. They are not only listening, they are paying attention and they are observing everything around them. Not only what the humans are doing, what the weather is doing, but what the plants and everything else is doing. Every part of this life is exchanging information.

The concepts are constant. They are consistent. Every human can go out there, into Nature, and they can experience the same thing. But everyone will have a different perception or have some different connection to It. These are the things that we hope opens up what we call insight.

And in the Native understanding, everything has what can be called Spirit, or it has an element of some kind of energy. When it has any kind of energy it is life connected.

A woodpecker occasionally rattles down a tree on the left. Cows moo in the distant fields. The angled but bright sunsetting light spotlights on bird level at the treetops. There they were, among the green leaves and branches, the waxwings, basking in the washing colors. And then an American goldfinch flows past fast. I didn't know the goldfinch was there until I saw him in front of me on my straight-ahead path, having just flown right by me.

The robin scurries up along the trunk of the fallen box elder tree, which overhangs at a 25-degree angle a wet field of buttercups and water grasses. . . . And then later, seemingly not knowing I was watching him in binoculars, he leaned over, pondered his flight, then executed it into a jungle of plants. How the birds must know the plants so well.

There must be a code embedded in the energy of a milkweed plant that drives its miraculously intricate and unique leaves, veins, shape, textures, colors, and the changing green-to-red forty flower buds on each stem. That unique code has no name. I want to feel that unique, no-name code for each species, each entity's unifying code that permits infinite configurations of a type. A raceme, an umbel, a daisy-like flower. The same for the birds and their traits encoded in the ethereal realm.

MIKE: How good it must feel when the birds sit on the dormant trees' branches and sing. Because their song doesn't just go out in the air as audible, it is also vibrational. It also goes through the birds, through their feet, right into the trees. The birds' songs are also actually helping activate the tree to come out of its dormancy. This is one kind of exchange of information I am referring to.

The concepts are coming right from the natural world, and this is why it is so key for people to spend more time outside. It is an exchange that you are benefiting from, and that is the purpose of why these things are out there. In one sense, they almost need the humans to bring that.

And it is sad that there are humans who actually think they can do a better job than Nature.

You don't have to be Native to apply these concepts. All you have to be is human and be able to differentiate this from that, not separate Nature from humanness. These concepts are for everybody. They are for all humans, for all life that's out there. . . . And the one that will never screw you over is right there. The wind, the trees, the animals, they don't have any agenda other than to enjoy this life. Unfortunately, humans know how to twist that.

There is a collective connection, such as how trees communicate with one another. For example, upwind trees send messages to downwind trees to change the chemistry of their leaves so that locusts or other dangers to the tree cannot take hold. It tells us that there is not only a collective in that they are always communicating, but also that they can convey a message that can help the ones downstream or downwind say, "Hey, if you don't change your ways you are going to get eaten up." It represents a higher function that you cannot get out of a book. I think they are retrievable by anybody who wants to dedicate enough time to allow the connection commitment to grow to be a major part of the life.

And those smartphones, etc. are interfering with people's abilities to function and not be influenced by these things. People are losing the ability to interact with other life forms.

MIKE: Philosophy and concepts are nice to read about, but if you are not living them then there is no connection to a philosophy or concept that asks, "Why am I doing this. Oh, because my superior told me this is the regimen for the day?" That's not life. Life is not mandated by laws that humans make. Life is mandated by the laws of Nature. The Natural Law has put them there, and you want to disregard that? You think that you're higher than the Natural Law? A couple hundred years ago, the same mentality was there. They said, "We can't have all these bison running around, we want trains running across this country. We have to clear the path for trains." That was a feeble excuse for the slaughter of the buffalo. You start to see there is no sacredness.

DENNIS: I heard on the radio news the other day that teenagers spend on average over nine hours a day in front of some screen—handheld device, TV, or computer. Literally it means that you're not looking at another person or you're not looking at a leaf or the ground or the weather. Screens, literally, physically, take people's attention away from the natural world.

MIKE: It's a form of isolation. But it's accepted voluntarily—these people are willing to do that. Just how much are they really missing? . . . I flash back to my earlier days walking up that little creek and that little gully coming from down the hills there in the countryside, flipping over stones, finding big salamanders and crayfish. I said to myself, what the heck is this creature, and you start finding new things. It's such a higher element of education that needs to be experienced by everyone. How do you mandate that?

JOHN: This is what's disrupting people's connections to the Creator. The Creator doesn't favor any religion. You want to meet the Creator? Just take a walk out there. That will explain who the Creator is. It's not a he or a she, it's an element you identify. There's intelligence right there. It doesn't come out of a book. It isn't coming out of a person's brain or mouth. This is a higher functioning part of life. This is what keeps this earth going.

MIKE: Where did the deviation of ignoring this Creation come from? It's not just money or control through industry and business. There is more: the disconnection from caring.

DENNIS: As part of some research, I've been asking students at the college and they've been asking other people if they are scared in Nature, scared in the woods and so forth. There seems to be a correlation between people who are afraid of Nature, whatever It is, and them wanting to build or have built environments. They seem not to be so concerned with the natural way of things.

MIKE: It gives them a sense of control or dominance. They say, for example, "We redirected that river, look what we put there, look what we can do."

DENNIS: It's a kind of megalomania of the human being thinking he or she is in control. Or can be in control. . . . They're taking a lot of drugs; they're taking a lot of chemicals. Something like 30 percent of the people are on antidepressants or antianxiety drugs. Some huge percentage.

MIKE: It is crazy. They're just masking the real symptoms and, again, I'm convinced it's because of the lack of understanding of the sacred.

DENNIS: Many people don't feel like they belong anywhere.

MIKE: No.

DENNIS: And when we walk, if I may say, when we go for a walk, we feel like we belong where we are. You can feel small and big at the same time. We know we're just teeny things, but we're connected to such a large thing such that you feel blessed and guided by this larger Thing. But if you don't have that, where do you belong? If you have family problems or school problems or money problems or whatever and there is not much spirituality or substantial support, then perhaps one is lost amid an alien natural world all around you from which you cannot find principles or purposes to live by.

JOHN: There is nothing to really grasp onto.

MIKE: So you start to see that the connectedness is sacred. That's the whole key. . . . Their concept is that new is better. They think that's an improvement.

DENNIS: Advanced technology is replacing the sacred.

MIKE: Indeed, it's replacing the sacred. For many, it's even damaging the thought process—there's very little intuition anymore. Nowadays, many people don't even know how to read a map anymore. "We'll just Google it." This is disconnecting more. Each generation seems to get disconnected further.

DENNIS: The newness and technology replace focus—but there must be a search for meaning in some fashion. Sacred is already there. You don't even have to search for it, you just have to be open to it. But with building over, building new places, it's almost as if you are dissatisfied with what is and you want more and more of something to replace the absence of something. And for many, Nature, and being alone—highlighted perhaps when one is in Nature without technological props—is scary. Scary means you are alone without much to protect you, little connection to a support system, a philosophy, a wide view one might not be able to see even if it is right in front of you.

MIKE: Why was this so important to make these technological structures? . . . Everyone talks about these great innovative technologies or systems that are in place. Well, because they deviated from a spiritual base and became so technical, they did themselves in. You better be careful; your system is vulnerable to collapse—it's happened many times in our past. We substituted nature-learning with politics, religion, finances, and Hollywood. These things become the distraction, instead of saying wow!

Nature and the birds are great elements that help us understand the meaning of life here.

DENNIS: Records indicate that most of the great thinkers and creators and artists and intellectuals and writers, and people in general, found their selves and their ideas when they were walking in Nature, especially the woods. . . . The inspiration of birds inspires us, and we learn about ourselves.

MIKE: It's the inspiration that comes from interacting with Nature. When you start to see things there, it should open other elements within the human and look at that and say wow! That really is amazing how this whole thing has such diversity. Yet it all works in unison, it all works together.

Nature doesn't divide people. They can all understand if we were to follow the natural ways. It's what Nature really wants. Humans think that they've created a higher function or concept that doesn't need Nature.

So, you can ride the canoe, but you have to know how to navigate. It is really important that the focal point stays in the natural world. That's not Native. The natural world isn't Native, that's the Creation for every living thing. But the White man, as an excuse, says, "We can abuse nature because nature is not our path, it's the Native's path."

DENNIS: So, if people are afraid of the salamander, for example, is it because they're afraid of the sacred? Are they afraid of the sacred or they just don't make the connection?

MIKE: Most have no connection to the sacredness of Nature and its members, salamanders, birds, and so forth. What they're afraid of is what's under the rock because you never know. . . . But it takes time to get a connection and say, "This is life." Back when we were very young, we did care about Nature, but we didn't know that the caring was the connection to the sacred. Until you get a little bit older and start to see connections, salamanders and crows and other entities might be scary.

DENNIS: Scariness means you're alone. If you are scared there is nothing there to protect you. But if it's sacred there is something greater there to connect with, to protect you.

MIKE: The goings-on of birds and other living things are cosmic events. Humans didn't invent them. Some human didn't sit down and write this out in his own script. It was not designed by a human being. We're supposed to follow what the Creation has put out there for us to follow. Most people will only look at it from their own limited perspective. For example, with the seasons, most people will focus on December 21 as the shortest day of the year. But from the larger picture, the shortest day of the year in this hemisphere is the longest day of the year in the southern hemisphere.

MIKE: The human scoundrels are doing very little but destroying Nature in their path, and they're the ones benefiting themselves. And they're keeping it a very closed group. Why do you think they do that? Because it favors them. These humans become wealthy, they become powerful, they get whatever they want. You don't ever see anything like that in nature.

JOHN: Right!

MIKE: This is why I think they have a vendetta against Nature. They want to destroy that because if people see the real message out here, that everything is equal and sacred, then they're done.

DENNIS: Nature shows them up for having a shoddy value system. They can colonize it and put Nature in a separate corner, on reserves and little parks where you can go visit once in a while.

MIKE: That's it. In the park setting, it's the way they want it to be represented to you, it's not in Nature's way. Even though the parks have to follow Nature's rules, the settings and the landscaping are not the way Nature made it.

DENNIS: Maybe even loneliness is another reason many shy away from really being in Nature. A lot of humans are lonely. Maybe Nature highlights human loneliness because they don't see connection in Nature, so they want to push it away. I mean perhaps they superficially see the interconnections of Nature, the animals, the plants, the birds, all really cooperating with one another. They only kill each other for food, right, and only when they must. We don't do that, many humans kill for fun or domination, or money. . . . So perhaps I don't want to look at that too much because it's going to show me up for having not as good of a system, as good of a value system. Nature has a higher morality than we do. "I don't want to go there cause it's going to point out to me that I'm a shallow, selfish person."

MIKE: Well not only that. Humans rank. What you have with the humans doesn't exist in Nature. In Nature, you won't get any recognition as far as your income, your status in the community, your intelligence, or your artificial intelligence. Increasingly, they follow that path of status and rank. That's why there has to be a big change. It's got to happen on that higher level. It's got to be that divine element. In Divinity, there's no Wall Street. There's no political agenda. The Divinity just goes. You want to see Divinity, sit next to that pine tree. Watch to see the birds and the squirrels and how they interact.

Mike summarizes John's practice of working with animals:

MIKE: John's practice of how he handles the animals is really his Native heritage. The animals, that is the highest—without using formal education or academia, the highest level of instruction. Because the animals are communicating with us, whether they are living or not. They are still holding and telling you, "This is who we are. We are the winged ones, and we bring medicine, we bring teaching, we bring pleasure, we bring ourselves for you to see us. You use our feathers; you eat some of us. Some of us are food." . . . We listen. . . . That's the blue jay we hear now [from nearby outside the kitchen table window]. Some are messengers. That is their duty, that is their role. That highest teaching, John has been interacting like that with animals for 40 years or more. At least.

There is a connection that he has. That when someone brings him a bird that was hit by a car, or was found in the woods, he lets that communication exchange. He lets that happen.

CHAPTER THREE

MESSAGES OF COLOR

Color links us with cosmic regions.

—PAUL KLEE

Humans have always been enlivened, enriched, and fascinated by color. Color is found naturally in crystals and gems, flora and fauna, sky and sea, as well as culturally, in dyes, clothes, ornamentation (including bird feathers), and in modern artificial pigmentation. Western ornithologists' discussions of bird coloration often center on protective coloration—for example, whether a species is ground-feeding, darker, or more colorful, suiting birds that spend most of their time in trees. Birds also use color and color patterns in preferential sexual selection for mating. Color is complex. Not only are there color variations on birds' different body parts, producing mutable patterns depending on viewpoint; bird colors change as the bird ages, especially from juvenile to adult, and as seasons change (usually bird color is brightest in spring).

Humans speaking the same language sometimes use color terms variably, or, with multicolored birds, emphasize different colors as descriptors. And there are multitudinous gradations of color, often unnamed in human languages. There are about eight hundred color terms in the Oxford English Dictionary, but these words describe only a small fraction of the thousands of color shades humans can differentiate. Moreover, different cultures and languages do not necessarily use the same (translatable) color terms or categories, even for some basic colors of the rainbow. An issue of concern for cultural anthropologists has been whether, and to what degree, different languages share the same, and the same number of, basic color terms.[1] Biologist Lee concludes that the "perception of color remains mysterious; it is as much a philosophical as psychological phenomenon."[2]

Whereas different human groups have the same color receptors in the eye, different cultures have varied understandings and perceptions of color:

> Many of the colors named in different cultures allow their members to discriminate among natural and useful objects in their environs, and different environments present quite different colors and color combinations. The Secoya . . . recognize green, yellow, and red, but not blue. The Hanunoo of the Phillipines . . . recognized only four colors black (*mabiru*), white (*malagti*), red (*marara*) and green (*malatuy*). Red and green represented other physical properties as much as color, red being dry and old and green being moist and fresh.[3]

Anthropologists and other scholars have conducted extensive research on the varying associations that different Indigenous peoples have with different colors. The Navaho, for example, associate color with the cardinal directions: East is white, South is blue, West is yellow, and South is black.[4] Pueblo Indians' cardinal directions are also associated with specific colors and species of birds.[5] Hamell speculates about historic Iroquois color preferences and meanings by tracing different colored beads in wampum and other beadwork. He notes that color in general is of great significance among these Indigenous people:

> Among the Northern Iroquoians, and the Northeastern Woodland Indians generally, color is a semantically organizing principle of ritual states-of-being and of ritual material culture. Three colors predominate: white, black and red. These colors organize ritual states-of-being into three contrastive and complementary sets: social states-of-being, asocial states-of-being

and anti-social states-of-being, respectively.... Within ritual contexts material culture functions to synesthetically manifest through its attribute of color the present state-of-being of its participants, and to synesthetically manifest through color, the desired state-of-being to be ritually effected.[6]

In *Bird Medicine*, Pritchard writes:

The color of a bird has meaning according to the Medicine Wheel teachings of each tradition, but these meanings change from village to village, or even person to person.[7]

The variations are endless, but the color of a bird's feathers are an integral part of the message.[8]

In this chapter, I present Mike and John's views on some basic bird plumage colors—yellow, orange, red, blue, brown, gray, and multicolored—without claiming that their views or preferences reflect those of any particular North American Indigenous group. One striking feature of Mike and John's views worth highlighting, and that seems consonant with those of many Indigenous people, is their unwillingness to cast judgment on bird color. They are interested in all bird species and color and are not inclined to associate particular plumage color with good or bad traits or omens. Their nonjudgmental stance—which entails a refusal to rank bird species—becomes apparent in the dialogue that follows.

COLOR AND SPIRIT

John and Mike read color not simply from the perspective of their personal preferences but also from the perspective of what Mike refers to as "the higher element." Color is a medium of connection to wider forces and sources.

DENNIS: So the variety of species, which of course also mean the variety of colors and patterns, are here in part to keep us balanced?

MIKE: They're all interconnected with the color and the characteristics that they reflect. When you get into the reds, like the cardinals, that's a different energy than the yellows and the blues. And you see that they are here to help us find that connection. Are they different for me than they are for John or you? Sure.

DENNIS: I remember, John, you talking about the junco, and the junco has silver on part of it, white and then gray. You were talking about how the white underside refers to purity and the other side on the back is gray, for carrying emotional baggage, as some people call it. I was wondering if you could tell me a little more about that. How is it that the junco has both those traits, purity and messiness, if that's the right word?

JOHN: That's just the balance. That's the way I look at it, anyhow. We always have to have a balance, no matter what we do. The reason why I say that about the junco is because that's what I get when I see him out here. The junco will come at different times when different people come here. It helps me to look out there and say, OK, this guy or this person has a lot of stuff going on.

DENNIS: If the junco comes?

JOHN: It's like yesterday when we had two nuthatches come. Now, with me, the nuthatches stand for faith. I don't know about anybody else. But for me it's proven that they stand for faith. And so, we had two of them come yesterday. Usually we only have one white-breasted nuthatch, and sometimes there might be a red-breasted one. So, there were two. And, you know, I had a difficult man here, with bone cancer, and we're supposed to do two ceremonies tonight, for him and another—a pipe ceremony, a crossover ceremony, for a man who recently died. So you know, I'm looking at this, and this is where I believe that the Creator or helpers come here and bring different animals to show us what's going on with the person.

The black-capped chickadee stands for truth, and you know this has been when my brother and my wife went up to Toronto, they wanted to find my grandmother's grave. Well, it's an unmarked grave, so I had a dream about where it was. So when they called me after they got there they said they thought they knew where it was. I asked is there this or this, or this there, referring to possible markers or signs that might point the way. They said yes. All of a sudden, the chickadees came to them up there. So my wife goes, John, you got chickadees right. I said that's the truth. . . .

DENNIS: Juncos carry a slate gray on their backs, which you said is symbolic of carrying emotional baggage like "a monkey on one's back." The white breast on birds' undersides means purity. What about Canada geese, which are multicolored, really? What do you think geese color and pattern mean?

JOHN: I know that the pattern is great for when they're on the nest. You can walk right by one.

MIKE: It is good for camouflage.

JOHN: You can walk right by one. That's why they keep their necks down when they're sitting on a nest. If another one calls to her, she'll put that neck way down and make it look almost as if it were a stick or a log.

DENNIS: As researchers have discovered, many birds can see in the ultraviolet light spectrum, which humans cannot. There is coloration going on for birds in their light spectrums that we humans can't see.

MIKE: Kestrels are one of those birds. They look for the trail of urine behind a mouse. Kestrels look for that refracting of the light or the ultraviolets that come off the refracting.

YELLOW

DENNIS: Every year I have a flock of cedar waxwings right near my house. They like the tall trees by the streamside. They're fascinating to watch. They're beautiful and elegant. They might be my favorite bird in terms of coloration. You know the waxwings have a little yellow tip on their tail.

Also, right at my house every spring and summer, are the yellow warblers, often around the crabapple tree in my front yard when it is in blossom and afterwards. Does yellow have a particular message? The yellow warblers are all yellow except for a little bit of orange on the tips of their breast feathers. Of course, there is the goldfinch, and others. And then also there is the Baltimore or northern oriole that comes around. He's an orangey guy of course, she's more yellow.

MIKE: There is a characteristic tied to the yellow, like the finches. If you watch their pattern of flight, they flap their wings, which raise them up, then they dip down, and then flap again. There's a rhythm to their flight that seems like they're always happy. Or they're always on a high note. You never hear or see them connected to a down, or a lower frequency. It seems like a higher energy, heightened, lighter.

DENNIS: I agree. They flip around a lot, kind of like the chickadees. I always thought about the chickadees that way, but now that you mention it, the yellow warblers and goldfinches are like that too.

MIKE: It seems to be the case with all the birds with yellow.

DENNIS: Now that I think about it, even the evening grosbeaks, which I haven't seen in a while, are upbeat also.

MIKE: And even the cedar waxwings, even though they're not a true yellow. They're in the shades of like a tan or close to a yellow. They seem to represent a certain frequency, or a certain level, that they seem content. I've never seen them at feeders or have any other interactions with any other birds.

JOHN: No.

DENNIS: John, you think the yellows are giving messages of happiness?

JOHN: Yes. Chief Skye and others I know always looked at them as brand-new beginnings, because that's what the color yellow

CEDAR WAXWING

Illustration by Julie Zickefoose

The cedar waxwing (*Bombycilla cedrorum*) is named for its silky soft feathers ("silkworm hair") and for its love of the small cones of the eastern red cedar tree. An elegant bird, some call it "sleek" or "handsome," with a "tailored look." Waxwing refers to the rich, red sealing wax look of the tips of its secondary wings. Overall, it has a muted fawn hue changing to light or lemon yellows. It also has a "rakish" black mask with a thin white border, making it look like a thief, albeit regal. Its complex regalia is topped off by a long, neat crest of blended feathers.

Other crested birds in the Eastern Woodlands are the cardinal, tufted titmouse, crested flycatcher, blue jay, and kingfisher.

With a friendly demeanor, sometimes cedar waxwings visit and circle near humans. Nonterritorial among themselves, they are very social; several often sit close to one another on a branch and pass food, like berries, from beak to beak without tasting any of it. Sometimes they groom one another. Their distinctive flight is one in which they often circle around the tree upon which they are perching, or a nearby tree, only to return a few moments later. They seem to play with one another. Sometimes they will eat overripe, fermenting berries and become intoxicated!

stands for. I think it also comes from the fact that the sun rises on the new day, in the east. In the [sweat] lodge, yellow stands for the direction of the east, and new beginnings. Unconditional love. Yellow stands for the sun, and yellow has always stood as east for me. We have ribbons out on the lodge, and they go yellow, red, black, and white.

(I've had guys come in and say that I have my colors wrong, because they come from out west, with the Lakota Way. So it would be black and white.)

DENNIS: Are you talking about the American goldfinch or just yellow in general?

JOHN: Well, the goldfinches. Because, like Mike was saying, when they fly, they flap up and down. That could be our human life. Instead of dropping way down and back up, we should be mellow with that. If you listen to the goldfinches, when they come back up from their little dips in the air, the bird will sing. Down, he's quiet. Back up, he sings. It's like he sings new beginnings, brand-new beginnings. So, it gives hope that this is how life should be every minute of our lives. We are going to have some little valleys, but they're not devastating. If you look at Nature, it's never devastating. That's how we look at it.

DENNIS: Beyond learning from others about yellow, this knowledge or this experience that you have, it's just there, and you pick it up?

JOHN: You just go with the feeling you get, whatever you are watching. It comes from something. It's not me. It's never me. It's always something else out there. It's always something else out there that's talking to you. That's how we look at it. . . . And for the cedar waxwing with the yellow on the tail, you know the tail works as a steer. That's how the bird steers and also how it brakes to slow down to get to new places. So again, it's a brand-new beginning. Anything to do with yellow—that's the way we look at it.

DENNIS: Waxwings also seem like happy birds.

JOHN: Really, and they hang out in groups, like family.

MIKE: All those birds that have that shade, or in that color range, seem to not have a lot of interaction with other birds. On the other hand, blue jays and robins, and a few other birds, will either chase others, or just don't seem to be like these other birds like cedar waxwings who stick to themselves. Chickadees, in contrast to the nonyellow birds, are similar in that they don't seem to be bothered by other birds around and will just sit there and stay in the company of blue jays if they come in.

JOHN: Back up!

MIKE: Give them space and back away.

DENNIS: Are you suggesting that happiness or contentedness is more likely when there's less interaction with other kinds of people or birds?

MIKE: It seems there's a higher or a better chance of happiness when you interact with the elements that are more in tune with yours. That's how they seem to function. They're harmonious within their own group. Whether they are cedar waxwings or finches, they all seem to maintain their happiness or a behavior that could be perceived by humans as being content. And that's maybe a lesson for people who want to be with others unlike them.

DENNIS: Humans need goldfinch and cedar waxwing therapy.

JOHN AND MIKE: Yes.

JOHN: You know, if you look at all the wildflowers, most of them are yellow.

DENNIS: Yes. The trout lily and coltsfoot and dandelions and asters and sunflowers and many more. Now's the season for the yellow birds to appear.

MIKE: It's almost like the yellow birds are having fun. Characteristics like when you were a kid, and you found something you could bounce on or jump on, and it makes you feel . . .

DENNIS: Light, literally light.

MIKE: So, there is energy connected to the characteristics and traits of the birds. When you see that and go "Wow," when they are flying, and you hear them sing, it just makes you feel better. Especially if you're having a little bit of a down day or if you have what some humans call "the blues." If you look at the birds that are blue, their energy is calm. When you are having the human blues it's a quieter time, even solemn. But I think in situations when the humans aren't so understanding of their own emotions, the blues can turn into depression and they don't know how to stop it, or they don't know how to redirect. The blues are good, I think.

DENNIS: It's when you're thinking more.

MIKE: Thinking more, yes. But if we're always at this higher mood like the goldfinch, if we're always like that, we don't have any reference to calming down or getting out of that and this, just being happy all the time. Maybe that's where ignorance is bliss.

Thoughts and attitudes are reinforced in Nature, when they confirm your suspicions. But the birds never leave those elements and go into depression. And we can be in our deeper thought, in our time of reflection. But we know that once we come out of reflection, we can be like the goldfinch. We can start to have a little more fun with each other and say, "See, that's how life should be."

This evening I saw two bright yellow warblers on half-dried branch stubs on a horizontal limb of an old white pine tree. They seemed positioned purposefully, moving about a foot or two away from one another, climbing around in such a way that their shape, color, and size, beautifully framed by the many little niches, dramatically offset the architecture of the limb.

A few minutes later, a cedar waxwing took off from another tall tree to my right and flew in a decreasing spiral, alighting on another tree only several feet away.

At the very top of the tree, still in setting sun, one waxwing of the small flock stopped nearby to look at me, then circled off with the others.

ORANGE

DENNIS: There are also the robins, which are partly orange, a dark orange, and there's also the northern oriole that's orange, very orange. They also both have black on them.

MIKE: It's like a charcoal.

DENNIS: The oriole also has a little white, too. But is the orange a yellowish color in the sense that you've been talking about it?

JOHN: I know with me, the orange always works as a warning.

DENNIS: A warning?

JOHN: Yes, that all started with what's called a red-bellied snake. . . . It's a snake whose body is pretty much a blackish color, but right behind the head it has this one red ring. When the snake goes to warm itself, it will go out and show that its belly is all red. And I remember out at the [sweat] lodge I went to reach for something, and this red-bellied snake went right up. I saw the red and I'm going, OK, why are you trying to warn me? Where I was reaching there was a hole and it had a bunch of bees in it and I had to reach right over that, so if that snake hadn't done that I probably would have got stung. And here I am, really allergic to bees.

So, I thanked him and everything. Since then—and that was a long time ago—anything that's orange usually I'll start looking for something that I need to pay a lot more attention to in my own surroundings.

DENNIS: Robins too?

JOHN: Yes, robins too.

RED

A red-tinged female cardinal happily preened herself by the many, fully open apple blossoms. Altogether she was a washed-pink scenario of comingling crimson, dark pink, and white petals and buds. A scene from heaven?

DENNIS: Then there's the reds, like cardinals. The flicker has a little red on him. The red-winged blackbird also has a little red. Is red different from orange for you?

JOHN: Yes.

DENNIS: The cardinal is maybe one of the brightest around, and the scarlet tanager.

MIKE: There's another one, the summer tanager. I saw one of those down in North Carolina. Boy, they are brilliant, they're much brighter than the cardinal.

DENNIS: And what's the red mean to you?

MIKE: I like seeing the birds with the reds. Mad Bear always used to say that if you see the red male cardinal in any of your walking or driving, and he crossed your path, it was a good omen. He said his appearance meant that there was good news coming. So maybe the red can get our attention, but it is not only a caution or warning. He used to think that it would bring good news, or you'll hear something good. Cardinals too have that demeanor such that they won't fly into the feeder if there are other birds there, depending on the birds. If they are chickadees and juncos, they'll come in, but if there are blue jays or starlings or other birds, they seem to keep a distance. I get the sense that there's an energy that the other birds pick up, as there were a bunch of blue jays that were out there a little while ago. They're still out there, back in the pines now.

DENNIS: The male cardinal is obviously very bright, although the female is beautifully and subtly colored. Cardinals also winter over. Is there a reason why the cardinal winters over, and is there a reason why the cardinal is red?

MIKE: I'm sure there is. Mad Bear used to talk about the cardinal only when the cardinal would fly across his path, wherever he would be traveling.

DENNIS: And that happens somewhat commonly, when the cardinal flies in front of you. . . . John, is that what you think?

JOHN: I never really have the cardinal come here. The only thing I was ever told was that the cardinal stands for the four directions. Someone said that the cardinal stands for the Catholic belief, but I think not. It just doesn't work that way. I don't think the Catholics were here before the cardinal, but that's how some people are. I have a friend whose spirit bird is the cardinal. When the cardinal does show up, I usually call him and make sure he's OK.

MIKE: You know it's Jim [pseudonym], it's his spirit bird.

JOHN: It usually works out right when the cardinal shows. I'll give him a call to make sure he's OK. And usually something is going on in his life right then and there.

DENNIS: So the cardinal tells you that there might be something amiss with Jim?

JOHN: Yes. Usually, you know. Usually when someone dies, you always find out what their spirit animal was, so that they can be remembered always, instead of having a stone monument there. That's our way of keeping that person always.

BLUE

DENNIS: What do you think about the blue jays?

MIKE: Well, we've talked about how the blue jays warn other birds and animals about humans and other potential dangers coming into their area. When at bird feeders, blue jays are pushy.... But the chickadees always hang out, they don't seem to be bothered when there is a blue jay at the feeder too. But other birds are reluctant to move in there. I think part of it is that blue jays have a demeanor that even though they may not peck the other bird, they don't give up any space: they won't relinquish their positions. Part of that is, you might say, is that they're not respectful. But it's a part of life and we say well, how do you want to be identified in life? What traits define you? The birds identify themselves in different ways—it's how they fly; it's how they interact. It doesn't mean that as humans we have the right not to like a blue jay. They have beautiful feathers, although their calling is not really pretty, and they can make a lot of strange sounds. I've heard them even imitate red-tailed hawks. But blue jays are pretty.

JOHN: As Mike said earlier, people always complain because the blue jay will knock the feed on the ground, but your juncos eat off the ground. So the jays are actually helping other birds and animals.

MIKE: As we have said before, people have their own ideas about the other life-forms out there. But when the human puts it into humanized versions, it's usually wrong.

JOHN: It's true.

MIKE: So, when you start to see from the perspective out there, you see that the intelligence still resides and lies out there. Of course, there's no system in place to measure intelligence in nature the way humans measure intelligence. So-called human intelligence measurements seem to be a major part of the human. The human wants to say, "Look how smart I am." But when I hear about all of the toxic issues out there, I think, "Are all our environmental problems because we're so damn smart?"

[Laughter]

MIKE: Nature is not arrogant and is more intelligent. So, you can see, from our perspective Nature is not only intelligent, but It has insight. I think human intelligence seems to negate or ignore insight. Some ask, "How'd you know that? How did you know this was going to go like that?" And most times, you don't know. But you do know that things are not going to follow the structure or the system the way things are handled in a laboratory. You're ignoring too many other factors. But they say yes, we're in control. I go, well, when you step out of that room you are not in control.

DENNIS: People get upset at the boisterous blue jays, and angrily think or say, "Blue jay, keep quiet!" There are not many blue birds. Of course, there's the eastern bluebird and the indigo bunting.

MIKE: The other species with blue are of a smaller stature. And they have characteristics more like a chickadee.

DENNIS: Right, bluebirds are kind of happy.

MIKE: They do say "the bluebird of happiness." . . . Regarding blue, there is also the great blue heron. They always keep their distance. You can never get close to them. Last week, I brought to John another injured great blue heron that got hit by a car right down the road from my place.

DENNIS: That's amazing that the heron had even got that close to the road.

MIKE: I know, it's very unusual.

DENNIS: It was dead?

JOHN: It had head and neck injuries. I didn't find anything unusual. The only thing that I keep saying is that the great blue heron has to do with "no more patience."

MIKE: Today's pace and influence is pushing everything to go faster. I think it's having an effect on the birds and what they have to tell us.

DENNIS: So you mean that the heron tells us not to be patient, but to hurry along. What with all the changes going on, that the herons no longer have the patience to behave the way they ordinarily behave?

JOHN: Most times with the heron, it means to move along with decisions and behavior, not to delay. Like with me and the human remains they found here in Attica. We went there, and I said we have to get them out. Well, it was twenty-eight days exactly, twenty-eight days for the next meeting. About a week or less before the meeting, I found a dead great blue heron. When we went to the meeting they started hum-hawing around, and I said no, the heron came, no more patience, we're walking out with the remains. It has to be, the heron said so. Sure enough, we walked out with the remains right after that. . . . And we buried the heron, of course. So the heron, with me, has always meant that I should not have more patience for a situation.

DENNIS: To move on past something that you're considering or waiting for?

JOHN: Yes.

DENNIS: So you are saying that this recent dead heron that Mike found is telling you or us to get going on something?

JOHN: Yes. One of the things that I was looking at was that I wanted to set up a booth at the school, on Treaty Day over in Canandaigua. I wanted to set it up about the residential schools and about RSS [residential school syndrome] that Native people got from the horrible residential schools that the government forced young Natives to attend. I've been kind of dragging my feet on it. Then Mike brought the heron over, and I said, OK, I guess it's time to do it, and everything fell right into place.

So, this is what I believe the heron was telling me. I think it's telling me more than that. Everything is falling right into place because I want to get some shrink to recognize residential school syndromes.

BROWN

DENNIS: Let me ask you about browns. There are a lot of different kinds of sparrows, most of them have either a lot, or some, a little brown. There's the field sparrow, the house sparrow, the tree sparrow, and so on. Do browns signify something to you?

JOHN: Most of the time they are telling me about the Earth Mother. Because most of the sparrows nest right on the ground. The field sparrows, swamp sparrows, and others will nest right on the ground. That's why it's the brown color, so it tells us that we need to stay focused on the Earth Mother and stay focused generally. Everything that we do we should be focused on. It's just like the whirligig beetle.

DENNIS: The ones on the water?

JOHN: Yes. They're constantly whirling around. If you look at them, they have a set of eyes that points downward into the water so that they can feed. But at the same time, they have another set of eyes pointing straight up, stating that no matter what you do, you need to always stay focused on the Creator. Stay focused on the Higher Sources. That's a real important teaching, because what we usually do is get into our own thing, do it, and never once think about anything else. We just have to do this one job and that's it. But we need to stay focused both ways. That's the way I look at it.

DENNIS: Nice. So, when you say the browns help us keep focused on the Earth Mother, does that also mean that they are helping us focus on the Creator?

JOHN: As I have learned from other Natives and for myself, for me the white on a bird always stands for the Creator. Sparrows always have some white. The brown thrasher has that brownish, reddish color. They are also telling you to stay almost hidden. To keep to yourself, not broadcast yourself, like "here I am" type of deal. To stay modest. Stay with the Earth Mother, stay with what you are supposed to stay with. Try not to go outside of that.

DENNIS: The cardinals and the brighter birds are in a way exclaiming their brightness, maybe their happiness. Is that the opposite of what you're saying? I'm trying to reconcile the two.

JOHN: There's always people who have to be focused on, like our elders. We have to focus on them. That's how I look at blue jays and the bright colored birds. When they're there, you must focus on them. When they're not there you get to focus on other things. If Mike were here and Bill [pseudonym] were here, my focus would be on Mike. You stay focused on them because that's what's important, that's where the teachings are from. That's where you know this is what we're supposed to be looking forward to.

MIKE: And the blue jays.

DENNIS: Mike, you should be wearing a cardinal outfit.

[Group laughter]

HOUSE SPARROW

Illustration by Julie Zickefoose

This ubiquitous, tough, adaptable bird (*Passer* commonly *domesticus*), a European import, once called the English sparrow, is probably the most seen wild bird, as it frequents cities as well as countryside (as do starlings). The whole population in North America descends from a few birds released in 1850 from New York City's Central Park.* The male house sparrow is easily recognized by its chestnut-brown back, gray cap, black bib, and ashy cheeks, while the female with a pale eye line is less marked in its streaky brown above and ashy white below. Along with several chattering calls, its common note is a loud *chirp*.

As it is brown, widespread, and bothersome to some non-Native human sensibilities, it is sometimes considered a pest. In England, sparrow clubs were formed during World War I to rid the countryside of them, and in China beginning in 1958 Mao Zedong had millions of house sparrows and the similar Eurasian tree sparrows assassinated. Ironically, their decimation led to famine in China a few years later because of the unchecked-by-sparrows increase in crop-killing insects.†

The house sparrow is sometimes unpopular with people who like only more brightly colorful birds: when some people talk about "bad" or "dirty" birds they include the house sparrow. Of course, Indigenous persons' perspectives and aesthetics respect this sparrow, like all sparrows, and birds in general. As John points out, all the brown-backed, ground-feeding birds remind us to honor the Earth Mother.

Familiar with people, house sparrows are somewhat tame, and they often ground feed on seeds in groups. As seedeaters, they have thick, conical beaks and special jaws, hard palates, and tongues to manipulate and open seeds. They like to dust-bathe in soil or bathe in standing water.

Blanchan wrote, "Even children who have never been out of the slums of great cities know at least this one bird, this ever-present nuisance, for he chirps and chatters as cheerfully in the reeking gutters as in the prettiest gardens; he hops with equal calm about the horse's feet and trolley cars in crowded city thoroughfares, as he does about flowery fields and quiet country lanes; he will pick at the overflow from garbage pails on the sidewalk in front of teeming tenements, and manure on the city pavements, with quite as much relish as he will eat the fresh, clean seed spilled by a canary, or cake-crumbs from my lady's hand."‡ Andrews writes, "The sparrow will show you how to survive."§

Considered "keen-witted," this bird thrives in habitats other species would find wanting, and breeds in all seasons. Pritchard relates: "One news report described house sparrows hovering in front of the electric eye sensors at the entrances to grocery stores, cafes, and other places where food might lie unguarded, causing the automatic doors to open and close so that they could steal food."** As with Mike's and John's perspectives and experiences, the appearance of specific species of birds act as messengers or reminders to reflect on, and take action with, one's own particular problems. Birds are helpers, spirit helpers, and personal therapists. They support the resolution of personal and social problems.

* Bull and Farrand Jr. (1977: 561). † See Todd (2012). ‡ Blanchan (1926: 113).
§ Andrews (1996: 191). ** Pritchard (2013: 194).

GRAY

The chatting gray catbird let me see him for just an instant before I had decided to give up trying to find him in the little fully leafed tree. He must have hidden behind all the while he was calling and singing song.

DENNIS: There's another bird that intrigues me. The gray cat-bird. The catbird, by human standards, is a noisy bird. Yes?

MIKE: Yes.

DENNIS: There are almost always one or two catbirds in the trees and high bushes next to my house every year. I'm clear that when I'm around the catbird is talking to me, squawking. Is there anything about the catbird that stands out to you? The catbird is also different because s/he is entirely gray except for a little dark crown patch.

JOHN: Also, underneath, on the rump, the catbird has a rust color. . . . We always looked at the catbird as having a double language. S/he usually tells us that we need to speak with a different language than we have been. That's usually what the catbird tells me, anyhow. What I mean by that is that you can talk like you're a truck driver down on Main Street and this bird will come and tell you to change your speech, stop swearing, and so forth. Sometimes the catbird means to change to a different ceremony. That's what I've always gotten out of the catbird. They almost always stay hidden where you can't really see them. You always have to look for it. Color-wise I never really paid close attention to it.

MIKE: I think there's a reason for the gray coloration like that. They stay more in the shadowy part of trees where they blend in. It's not a real dark gray, nor a light gray. The shade of gray matches the colors when you look under the trees and see the different shades of the shadows that match their camouflage.

It's rare for them to come out into the open. I know when John and I were over at that one property, I was walking the property line and I was burning sage and tobacco. This friend of mine and this lawyer were having trouble with their neighbors, so we walked the line, and I had to make it clear that the tobacco or sage wasn't taking sides. There was a catbird that followed us around the whole perimeter. When I finished, the catbird came out into the open. But he was not squawking. He wasn't making his call, but he walked. He followed us through the whole walk around that boundary line.

DENNIS: The catbird seems to like to be around humans.

MIKE: They do. But they don't want to get too close or be seen.

DENNIS: John, if I understand you correctly, are you saying that when the catbird is squawking near a human s/he's trying to tell the human to change the language being used?

JOHN: Like you need to look at the language.

DENNIS: It's a way of the bird telling me that I need to look at myself a little bit more, and change the way that I'm talking, because I'm not quite on the spot, or doing it the right way?

JOHN: Or maybe you're offending other people. The bird is trying to help you become a better person.

DENNIS: Like a counselor. . . . The catbird is a teacher. It may not be a teacher for every person. John, you were never taught that the catbird was a messenger about language, you just came to understand that on your own?

JOHN: So what happened was, every time that I went to Chief Skye's house or to this other man's house, the catbird would always show up beforehand. I know that catbirds have a nice song as well as that cat call. I figured that Chief Skye used to speak the language. The other man would speak Seneca, but it was a newer version of the Seneca language. So, I started to look at that and said to myself that maybe the catbird has to do with the language. Maybe this bird is really starting to tell me to look at the language. So, I started looking at the Seneca language, and then I also started to realize how much I was swearing.

DENNIS: So, it's not like there was a story handed down over generations about catbirds' messages?

MIKE: We're in a very interesting time. The Natural system is continuously still trying to help. I guess from a human perspective you'd ask, "Why are you trying to help these idiots who are trying to destroy you or cause harm to you?" But the Natural system can't pick and choose. It has to go for the good every day, to support the life. Whether that life understands or not.

DENNIS: Even if John's the only person who understands the catbird's message, it's still the catbird's duty to inform. Is there an ability about the catbird, or all the birds, that despite the circumstances that they find themselves in, they continue to try to teach? Or try to live the life that is meant to be, even if very few humans around them recognize it?

JOHN: We base our life on everything that is out there. That's what we should be basing our life on. Very few people do. Everything is out there as a teacher, whether you want to look at it or not.

MIKE: It's the proof that the Natural system can only do good. It doesn't say, well, for this group of people it gets this kind of bird. It gives it all to everybody. It doesn't read into who's trying to seek it out, and say no, they're not ready. If the humans are ready, then they will recognize it. But if humans keep behaving like they're not ready, then that message and system will remain elusive. It will not be recognized.

This is difficult. This is why you try to give referrals or when some of the kids ask, are there other books I can read? There is no quick read or fix. It's still not going to make the connection for most people because they've been so indoctrinated by man's system that says we have a book, or a website, for everything.

MULTICOLOR

Chickadees in front of, and within, the young willow bushes. One definitely came out to the end of the shrub to exchange greetings with me, while two

or three others hopped about out of sight. He was like the scout, who after
several chickadee-dee-dees, returned to his family and friends. Later
he appeared in the sumac garden closer to the house.

Several minutes later walking back to the house, I stopped in my
tracks, remarking in my mind, and out loud, that the encounter with
the chickadees in the freshly light-green willow bushes was wonderful.
That experience back there, in time, in space, in place, but still fueling
my thoughts and feelings. . . . How long do you smile for beauty?

DENNIS: The chickadee, the black-capped chickadee, that we
have around these parts, is multicolored, with shades of black,
gray, white, and brown. Can you talk about the multicolored-
ness of the chickadee?

JOHN: It's really kind of special. Especially with the black eyes,
and the black around the eyes. Because the chickadee is telling you that you never know where the chickadee is looking.
Because its black eye has a black cap background. That's
important because you don't know if the bird is looking at
you, or someplace else. The white part, of course, is on the
belly, and the gray parts are on the back. The chickadee is like
the junco, very similar. But the juncos are different because
the juncos always seem to be so heavy weighted, they're always
on the ground feeding, whereas the chickadees are always up
on trees and often they're hanging upside down. Like you said,
it seems like the chickadees are constantly playing, as if life is
so good, they get joy no matter what.

DENNIS: They are fun to be around.

JOHN: They're just enjoying everything. I think that's one of their
biggest teachings, that we have to enjoy life. Also, especially
nowadays, in line with the black around their eyes, we have
to try to make sure that people don't know exactly where we
are looking. Because, if we might not fit into others' views,
into the rest of society, you almost have to hide, sometimes,
your inner thoughts and views, not to let your full face, as it
were, be seen.

BLACK-CAPPED CHICKADEE

Illustration by Julie Zickefoose

This common bird (*Parus atricapillus*), sometimes called the eastern black-capped chickadee, is one among other chickadees in the United States and Canada, east and west. Smaller than a sparrow, it is plump, with gray, black, and white markings. Like all chickadees, and the tufted titmouse in the same genus, it always seems very busy and sometimes hangs upside down in search for insects. Its acrobatics are not disturbed much by humans nearby, and it appears tame, like the chipping sparrow, perhaps the tamest of all local species. Chickadees like human company, and are generally inquisitive, often following humans on walks or meanderings in field and forest. Their black cap (Latin species name *atra-*, black, and *capill-*, hair), black bib, and white cheeks make them quite noticeable, as does their cheery onomatopoetic call *chick-a-dee-dee-dee* or *dee-dee-dee*. In spring, they have a distinct whistle of *fee-bee* or *fee-bee-bee*. They have several other shorter notes and often chatter. You can imitate or whistle their call and they may appear out of the forest to examine their mimic.

Chickadees often flit about the branches of a tree and usually travel in small groups, playful and sociable among themselves. They form flocks of about three to twelve individuals in winter that contain some full-time regular members and also some floaters who move around between flocks. If close by, you hear a faint rustling of wings as one flits about. They can also roam with other species such as nuthatches and woodpeckers. Fearless, "brave" as John points out, they are hardy and spunky and usually not threatened by other birds. They can be seen all year round, usually in woods or the edges of woods, especially conifers. The black-capped chickadee also likes willow and poplar trees. Chapman calls these birds "merry little black and white midgets."*

Andrews reports that for Cherokee, the chickadee is the bird of truth, helping us to pinpoint truth and knowledge. He also suggests that the cap refers to thinking, higher perceptions, and mystery. The cap's blackness suggests that the chickadee can "help you with the uncovering of mysteries of the mind . . . and higher truth," and "help you to perceive more clearly in the dark."† John says that the black eyes in front of the black cap, preventing humans from seeing where the chickadee is looking, tell us to be constantly aware, and on the lookout for false fronts.

* Chapman (1913: 179). † Andrews (1996: 126).

DENNIS: Chickadees' "hiding" where they look is a lesson to us humans to be the same, to hide where we are looking?

JOHN: It's like the barred owl. We call them "dark eyes" because they almost have a black eye, although it's really a very, very deep blue. With owls, you can almost always tell where they are looking, but when it comes to "dark eyes," the barred owl, you can never tell. You can never ever tell where the barred owl is looking. It tells us that you always have to watch what you are doing because you are being watched whether you think that birds are looking at you or not. It's the same thing as the chickadees. The chickadees are constantly watching us, so you know you have to watch your behavior.

Mike, what do you think?

MIKE: Well, that's it. If you think you walk outside and you are not being observed or watched, you're not paying attention. They are all watching.

DENNIS: The cool thing about the chickadee is that s/he's such an upbeat bird. S/he's such a joyous bird, a lighthearted bird. But at the same time, it's giving this serious message. Usually, I don't think of light-heartedness and seriousness combined.

JOHN: It is like that sometimes with Native elders. Chief Skye did that with me a couple of times. He would put you in your place and be so nice about it. But the whole time you are thinking, I just got my ass scolded and this guy has a smile on his face.

[Laughter]

DENNIS: Some people I would say, even the three of us around this table, when we talk, we can be both lighthearted and very serious in the same moment or in the next sentence. Also, sometimes we laugh about tragedy, or we cry about comedy. You can combine these emotions. I think people who are mature, or whose perspectives are mature, can feel both at the same time.

Alright. So, the chickadee, the black-capped chickadee, is a very happy bird often following us around. He or she doesn't seem to have the fear or wariness that other birds and animals have. Maybe it's not fear, but carefulness. Why is that?

MIKE: As we said earlier, the chickadee seems to be of a different character in the bird kingdom. They seem to be very harmonious with all, even all the other birds. They're very tolerant of any kind of bird that comes into a feeder, and they don't run away. They are not aggressors. I've heard people refer to them as being very brave. You don't have to be an aggressor to be brave. You just stand and hold your ground and chickadees seem to be like that. Maybe there is a lesson there too.

CHAPTER FOUR

NATURAL LAW AND ORIGINAL INSTRUCTIONS

Consequences, Changes, Connections

Birds are indicators of the environment.
If they are in trouble, we know we'll soon be in trouble.
—ROGER TORY PETERSON

It's an accepted fact that human-caused climate change, along with deforestation, habitat destruction, pesticide and herbicide use, and animal poaching have significantly contributed to the unprecedented decline and destruction of bird and other animal populations, and to biodiversity decline in general. Keesing et al.'s landmark study in the journal *Nature* concludes that as diversity decreases, the balance of nature is upset, and remaining species are even more vulnerable to human influence and more likely to spread powerful pathogens.[1] Approximately three billion birds, about 29 percent of the bird population, has been lost in North America since 1970.[2]

Over the thirty years I have been watching and interacting with birds around my homestead in upstate western New York, I have personally observed the reduction or disappearance of numerous bird species, including the decline of white and red-breasted nuthatches, bobolinks, cedar waxwings, barn swallows, orchard and Baltimore orioles, ruby-throated hummingbirds, various warblers, and the complete disappearance of evening grosbeaks, eastern meadowlarks, ringed-neck pheasants, wood-cocks, and mockingbirds.

In what follows, Mike and John discuss their understanding of Natural Law in the context of their observations of recent changes in bird behavior in this Eastern Woodlands area.

DENNIS: We have talked about Natural Law, a concept closely related to the phrase "Original Instructions," which was made famous in the book edited by Melissa K. Nelson.[3] Can you be specific about what you mean by Original Instructions?

MIKE: My reference is to Mad Bear and what he meant by Original Instructions. When bird species live their lives according to their own characteristics, they are demonstrating and following Original Instructions. They have not changed them. They have not adapted them to be different. Their feather color, their call, their way of living their lives, whether it is a blue jay, a robin, or any of the other numerous species. That is one way of defining Original Instructions.

JOHN: How does the change of a bird's habit fit into that?

MIKE: They can change. Some of them can change.

JOHN: Especially lately.

MIKE: But the basic elements of what we are familiar with, of how a blue jay or a cardinal or a robin lives its life, usually falls in line with what we have observed over our lifespan, or our generation of now. But there are accounts that have been handed down from previous generations that will describe ravens, describe crows, and talk about birds of prey. Which we can now say, they are following their Original Instructions. Now we can see the previous generations descriptions demonstrated.

DENNIS: So that includes things like what they eat, how they mate, what kind of niche they have in the forest, etc. The day-to-day details of what they do.

MIKE: Absolutely. Those show us clearly how to define Original Instructions.

DENNIS: So, all we have to do is look at a cardinal and we know that that is the way it is supposed to be?

MIKE: This is how they were in previous generations. And this is how they still are. . . . They still get red feathers, they still go through a stage of immaturity where the colors are not defined, and then they reach maturity. Now we can know how they live their lives, and the foods they eat, the calls that they have, which falls in line with the historical knowledge.

DENNIS: So how did the cardinal get its Original Instructions? From whom or what?

MIKE: There are legends of the birds, and how they got their songs. Earlier I told you the story of the eagle and the hermit thrush, and how the thrush got its beautiful voice.

That is confirmation to me. There was some structure, some interaction, behind whatever you want to refer to as the Element that assigned Instructions, or maybe they did not really assign them, but They said, this is how this will work. Let's put it in terms that a human could try to understand. These terms exist in Creation, or in Nature, but they are so difficult to understand, so let's put in the guise, or circumstance, of a competition for survival. How these birds all reached whatever level they have achieved since what we would call the start of Creation.

DENNIS: Would it be appropriate to say that the Creator gave the instructions?

MIKE: We can make that leap. But some people are hesitant to assign authority like that to one being.

DENNIS: John, what would you say?

JOHN: Leave it alone. I wouldn't say anything, to be honest with you. I really wouldn't.

DENNIS: But you recognize that there is such a thing as Original Instructions?

JOHN: Oh yes, but it is like trying to study the Creator, whether it is male or female, things like that. The way I look at it is, just accept it.

MIKE: Humans like to anthropomorphize. We will assign the Creator an image of a human. And then by looking at ourselves and others, we say, "That's quite a task. We have to look at the habitat that these birds function in, and have to look at each bird, and assign a habitat." You can see that is not a very probable scenario. If you want to accept that as your version of how birds got their characteristics, it's OK. But I just can't do that.

DENNIS: So you don't use a word or sentence to describe the forces, or entities, or whatever it is that laid out the instructions? You are basically saying that we are using the word "Creator" just so we can keep talking. It sounds like you more or less agree with John, but it is just that for the purposes of discussion, it is OK to use the word Creator.

MIKE: We need to go beyond the human views and concepts. The human makes this up and says, "We want to assign everything that is in Creation to the Creator. That the Creator had the directions for everything. Here's the directions for this bug, here's the directions for this water." . . . And so forth. Really? We are really thinking like a human to assign that to a much greater Intellect. We don't know how to describe it. Do we want to call that Intellect, or Intelligence? Well, I think that whatever that is, it is much greater than Intelligence. Much greater. For academia or intellectuals to try to condense it down to that level, you are not understanding what this Creation can do. It is not regimented by the presumption of what a human thinks. That doesn't even apply here. So to take that position really has narrowed your ability.

DENNIS: To define It itself, is a violation of the process.

MIKE: Right, and I have even heard elders say that for someone to try to define, to describe, what the Creator is, you have already lost that ability to understand.

So you take those words, and you go, "Wow. Heck, how did they know that?" Well, because that is what was handed down to them. And did people say, "Let's take this to council, or let's document it in a hearing, or let's establish what are our findings about the Creator?" No. They did not function that way. They never did.

DENNIS: You cannot presume to use words to describe it. To understand it.

MIKE: Right, we really can't. . . . And I would ask astrophysicists why is that a locust tree? And ten feet away is a pine tree. And they are using all the same elements. But why is one this species, that is a pine, then there is an apple. . . . You see how complex things can get if you start looking from the analytical, or academic, perspective. Because the academic wants to say, "This is how it is, and we can prove that." . . . If that works for you, OK, but do not assign it to the traditional teachings of our ancestors. I do not know how their minds worked, but they knew how to hand down the messages.

Do not be misled, do not take some changes to the animal behavior as deviant, or they are not following instructions. They are adapting. Does it exist, that they have to adapt? Of course. Everything that is living in this Creation wants to continue to live. In order to survive, they are finding ways to follow their Original Instructions.

DENNIS: John, would you say that now that the red-tailed hawks are partly scavengers, is that part of their Original Instructions?

JOHN: Yes, because they have to survive, and just like Mike was saying, they have to adapt.

DENNIS: So the Original Instructions could include a range of potential behaviors that might be necessary for survival, but perhaps could not be understood ahead of time?

MIKE: It was not foreseen that DDT would lower their numbers, to the point that they might go away completely. But it got to the point where its use had to be stopped. Fifty-five to sixty years ago, when I was a kid, there were hardly any red-tails around. And now that DDT is not being used, I see them every day, everywhere, sitting on a telephone pole.

DENNIS: John, do you see Original Instructions in your rehabilitation and care of animals that you are bringing back to health?

JOHN: Yes. They act one way. The way they are supposed to act. They do not know; they do not understand that I am there to help. And sometimes, especially when they start feeling better. . . . You know I have had a great horned owl: you hang on with one hand, you open the beak, you put the mouse down, and that bird will not bite you, until it feels better. But all of a sudden, go out and try, and the bird will nail you. So, the instructions are that these animals are meant to be free to live their lives the way they are supposed to.

MIKE: It is basically saying that they do not need anybody's help. When they are feeling healthy, and doing what they are supposed to, they don't want your help.

JOHN: Yes.

MIKE: So, John, are you suggesting that you are forcing these animals to get better?

JOHN: Oh, yes. But the way I look at it, is whatever animal comes in, that is my teaching about that animal. I have to learn from that animal. You just do not take an animal in and say, "OK, I have this bird," and then don't learn anything from it. Because the bird is going to tell you what you need to look at. Just like that woman's owl. The owl is telling her what she needs to look at. Whether it is dead, or whether it is alive. It is still telling you.

DENNIS: When a dragonfly flies around a pond and mates in midair, that's part of its instructions. So, it is what it is? And there is not much more you can say?

MIKE: Right, and Nature will continue to teach you. Just because it taught you one version of instructions does not mean that that is absolute and that it won't ever change. Sometimes they need to make those adaptations, need to evolve, need to change.

DENNIS: Are you saying that the ability to change, to adapt, was part of the instructions? It may not have been specific at the time, but it was built in so that they could change to adapt?

MIKE: There could have been a little sub-note put in. Humans would want to say, "I didn't read the fine print."

[Laughter]

But if your environment should change dramatically, now you still have to survive. This is built in. But if the environment stays healthy, balanced, and continues, you do not have to access that component, that part that has been given you. Unless it is absolutely necessary. So it makes sense like that, when we are pressed into a situation.

Mike relayed how he had seen two downy woodpeckers under his back porch and that they were picking larva out of a mud wasp nest. He had never seen that kind of behavior before. John agreed that he had never seen downy woodpeckers eating that kind of food.

JOHN: The woodpeckers would also get spiders because the wasps also pack spiders into their nests, an unusual foraging method for them.

MIKE: I had never seen them do it any other year, even though we always get those wasps in there. It is like how herons now have insects in their stomachs. . . . John, was it a screech owl also eating insects?

JOHN: We had a screech owl, and we had two hawks with insects in their stomachs. These animals that normally would not be eating insects are now eating insects. And now we have catbirds and robins eating maggots.

MIKE: With these birds not eating the foods they're supposed to be eating, it makes you think: what the heck is going on that's interrupting food supplies to the level where these animals are literally starving to death? And there's food around, but the foods they're supposed to have are in such low supply that they're eating alternative foods. I'm going to go out on a limb and say that eating the alternative foods are most likely causing them to function at a lower level, and that's why they're losing their lives. When you start to see humans interfering more with Nature, the greater the impact seems to be.

JOHN: So, when we see the birds not eating normal foods, is that part of that what you are talking about, that we're seeing a major change?

MIKE: There's a shift because the changes have already started. The shift has begun. I think the shift is in constant motion.

DENNIS: It is clear from Darwinian studies that the shapes of birds' beaks evolved over time to adapt to niches or particular kinds of food. Like the nuthatch can climb upside down and get the insects that way. Not many other birds can do it that way. So, they have that little niche behind the bark because they climb down the tree upside down. They have evolved over time to fit. So now, if there is a different food source that a bird is starting to use, in the long run perhaps the bird's physique may change to adapt to that niche.

JOHN: Whether it's the feet, or whether it's the beaks. . . . I like what Seneca Chief Skye says, "I don't have to wear my eagle feathers on the outside, I keep them on the inside." I always liked that from him.

DENNIS: The task is not so huge that way too, so it's not like a burden.

JOHN: Yes, exactly.

MIKE: You think the climate changes are hard on us? The trees, the plants, the animals, they're feeling it, too. Those hot days hit, and I see the crows sitting in the shade with their wings out and mouths panting. They are looking for a breeze to cool them down. It's brutal. Just the other day, I heard the crows making a ruckus and then I heard a raven. The crows chased the raven. I didn't think they would. But the raven started doing that guttural call, the deep *Urr*, and then he flew away, and they actually chased him. And then they flew back, and I said I never knew crows to do that to a raven. I mean they are all in the same family. So it was strange, and I never saw it before. But times are changing.

Regarding a tree's strange fungus-like emanation that Mike did not recognize:

MIKE: I've never seen anything like it before. There are a lot of things now that I've never seen before. It's getting to a stage where these anomalies or unusual things seem to be getting more and more frequent. It's very unusual because Nature typically doesn't allow foreign things to just manifest. . . . Mad Bear said that in the times of change, or what the Bible calls "end times," there would be strange bugs and beetles and other anomalies that would show that it was a time when the change was going to happen. I get a sense that these anomalies are caused by the humans and their technology. . . . We were up in Lewiston over the weekend for the Harvest Festival and there were many turkey vultures. Probably forty or fifty, all circling, and they were all hanging around this cell tower. And they landed on it.

DENNIS: Do you find the same thing, John? Are you noticing more things out there that you've never seen before in your whole life? I mean not just the turtles, but the plants, the birds, and so forth?

JOHN: Everything. Everything's just changing, and it's changing fast.

DENNIS: So that's good and bad, right?

JOHN: It's just bad. That's what I think.

MIKE: There are some alarming things that are happening. If we do not start to make some changes to allow things to return to normal, or restore themselves to their more natural habits, I think we are going to see an environmental collapse. It may not be total, but it will be devastating.

JOHN: Well, it has got to be really close now. If you start to look at everything and look at where the insects are. How can it stay in balance? It just can't. It just can't stay within balance.

MIKE: I guess if there's any good in these changes, it is that they are indicators of what we're heading into. I think for the purification to take complete manifestation it needs to get to that point, because the humans don't seem to care.

Life was given not just the opportunity, it was given the ability to say, "Well, the Creation story says it doesn't want life to stop." So you have to keep this whether you are aware of it or not. Therefore, I say—or the elements that move through prophecy say—you're going to get enlightened whether you want it or not. Your definition of enlightenment is whatever you want it to be, but I think it integrates the cosmic part of life to the human. It says if you're going to reject it, if you're going to just turn us down, then join the past civilizations, because that's what happens. Like in Egypt—the people who built these structures are gone. Why? They got arrogant. They thought they were way beyond the average human, and they were so selective about what life-forms they wanted to have around. That's not the cosmic way. The cosmos does not discriminate or tell you a particular life-form doesn't deserve to be here.

This ties in not just to prophecy, but the changes in which things are going to improve. It's going to be on Nature's terms. It's up to the humans if they want to make changes in accordance with Natural Law. But if humans are still reluctant, then Nature is going to say, "If you want Us to make the change then We'll do it, but you no longer have any involvement." The humans have lost respect.

JOHN: That's right.

MIKE: Whatever is happening at the time, you can go Wow! This earth really is a living element. It wants to provide for every living thing out there. It's not selective in saying this year we're only going to cater to the seed eaters or next year we're just going to provide for the ones that eat worms or insects. It's this whole element that says it's here for everybody. Most humans have lost their ability to understand this. I think part of that is because politics is such a distraction. It alters brain function. Religion does the same. Finances and politics and religion are the things that secure their interests. They don't look beyond for the next generation. They're not looking to say, "What about our kids? Are we looking that far down?" No, they say, "Why should we? Take what you can now."

And that's why we're in the situations that we're in. It is unfortunate that we have to learn the hard way. We need to return to more natural methods of getting food and medicine, to pay more attention to natural cycles.

MIKE: So, when there is an equinox, you should be welcoming that energy to come back out of the ground and manifest as the medicines and the foods that we need. Most people don't make a connection to the equinox. But if we're going to harvest medicines that are in root form, if that's the source of the medicine, then we should harvest them before the equinox, because that energy is still concentrating in roots and under the ground. This is the kind of concept from which people have drifted. Organized religions have stigmatized the Indigenous type of belief system. Yet the Indigenous belief system followed and recognized the structure of the Creation which we call sacred.

DENNIS: So we're doing what we do, I guess, with respect to the prophecy. We're doing this knowing that the vast majority of people are not paying attention. So what we're doing, if I may include ourselves, is to help the transition so there are more people around who have a strong sense of connection, so that the transition will be smoother. So we're trying to educate others, though we know that many people are uneducable in these regards.

MIKE: Yes.

JOHN: With prophecies, there's so much that isn't talked about that we're being taught, even just about how to be a better person. You get lessons that come through the [sweat] lodge. . . . People go "yeah," skeptically. But then a year later, they're actually walking what they picked up in the lodge. The lodge is for anybody who wants to look at themselves, to be able to get rid of the nonsense, to look at how it should be according to Natural Law.

DENNIS: What is Natural Law?

JOHN: Natural Law is, the way I look at it, is where everything has to work together. To me, it is just that simple. I do not break stuff down, like everyone else does. . . . My way is acceptance. I accept that, like Mike says, the locust tree that grows next to the box elder or the pine. I accept that. I accept that because we are supposed to accept other humans. That is what I believe. Natural Law is basically just acceptance of how we should be. We should be the best that we can be. Like Mike has talked about the blue jay and the robin. The robin is not trying to be a blue jay, and the blue jay is not trying to be a robin. But the blue jay is the best blue jay it can be. And the same thing with the robin. . . . Those type of teachings to me, that is Natural Law. We have to try to get ourselves to being Natural Law.

DENNIS: It sounds like, in part at least, that Natural Law is harmony, getting along. . . .

JOHN: How about surviving in a very good way. And being connected to the Earth Mother. It is nice and simple.

DENNIS: It's harmony between things.

JOHN: Uh-huh.

DENNIS: So, the system is harmonious. And that is the natural way of things?

MIKE: It is how Nature functions.

DENNIS: A lot of White people think it's about humans against Nature. Conquering Nature. They like their four-wheelers to make a lot of noise in the woods, and they display themselves as if to show how great persons they are. Is it a violation of Natural Law that some humans are destroying the environment?

JOHN: Sure.

MIKE: Yes, and there are consequences. They do not believe that. But the groups and the individuals that do those practices will find, when they leave here, they will be given insights into the impact that they had on other lives. They will get to feel what they did to that environment. As a consequence, not a punishment. To help them understand if they were to return, they can never claim, "I never knew."

DENNIS: So such humans have violated their own Original Instructions?

MIKE: The instructions to human beings are consistent across the board, with all the cultures, of what they were told, and how to be, not just with the earth, but how they should treat the animals, how they should treat the environment. They were given those instructions, those insights. You do not want to be a part of the group of people who hurt or damage the environment, not only the earth, but the life that resides in each one of these habitats.

DENNIS: Is it correct to say that the hawk, who is not violating his Original Instructions by eating roadkill, contrasts with the humans who violate their Original Instructions by chemicalizing the planet? Animals cannot violate their Original Instructions. It's as simple as that?

MIKE: Right. Because the humans were given a form of free will, that basically said this is how you are supposed to live life. But you are not forced to. Because you are being gifted free will that says you can violate your Original Instructions.

DENNIS: So the system, if you will, was set up imperfectly, because it allowed humans to violate Natural Law.

MIKE: The way I have heard it stated was that if the humans want to truly show their love for the Creation and the Creator they will listen and abide by these instructions. . . . Now let's humanize again—how can you say that the Creator says, "I am going to make humans, and they are going to love me." And you think about it, and you say, "What kind of love is that? You have no choice?" Behavior says more than words.

The prophecy is a warning. I compare it to when you are driving on a road you've never driven before, and you see a bend in the arrow on the road sign. If you don't slow down—and that bend is sharp—you are going to go off the road. You're going to have an accident—but who do you want to blame? Everybody wants to blame somebody.

JOHN: This is what I'm saying about the teachings of birds, of Nature, of the sweat lodge. It can come through these sources about our lives as we know them, how we've been taught or programmed. Right now, we're at that bend. Do you want to drive too fast and fly off the road, or do you want to look at the changes you need to make? To understand, respect, how it fits into your life, and everything else that goes on. Or are you going to be part of the great separation that separates you from Natural Law?

MIKE: But that's the thing. As much as humans are abusing the earth and the natural world, the earth is still helping the humans who want to understand, to get the help from that system. The birds are still giving us messages and lessons. It shows you that not only is there a much higher intent, but there is a higher function. The earth doesn't hold a grudge. It doesn't say, "Well you didn't listen to me for the last forty years, and now you're going to come to me." The earth doesn't care: the earth still provides, whereas a human might shut down and say, "Screw you, you've been digging holes in me."

DENNIS: I thought climate change was a response.

MIKE: It is, but the earth continues to provide all of the elements that we need. And sometimes it's on an unseen level.

JOHN: Yes.

MIKE: And you still will hold the same concepts, still keep doing what you've always done because you know that's what you have to do.

JOHN: Yes, those are your values. You have to do what you have to do.

DENNIS: Is that why you do these things, because even though you know that as prophecy says, that things are going to get worse before they get better, you try to ease it a little bit?

JOHN: What I tell everybody in the group is that you know when it comes to the environmental end of things, and fighting for it, we are spitting on a fire. You have to understand this so that you don't get way off in left field getting all mad. We're doing what we can to save what we can, and that's all there is to it. That's why I won't get involved with the pipeline—someone else can take care of that. I'm taking care of a little piece of Earth Mother here in this county. That's all I'm doing. One stream at a time. Are we being met with resistance all the time? Of course, we are.

DENNIS: But that's what needs to be done, to spit. To live a life, you have to spit on the fire. If you are living a life with integrity, even if the fire gets bigger, you have to spit.

MIKE: You start to think about how do we stop feeding the fire to the point where it's causing the damage? Can we all be fire keepers or managers? Yes, and we should be. Don't keep escalating the fuel. Start realizing that it's harmful to every part of life.

Each time the world has gone through drastic changes, the life that has remained has become higher functioning. One of the concepts that I get from that, as I said earlier, is that you are going to get enlightened whether you want it or not. I like that because it says that this cosmos wants to get us to a higher-level system.

But we can achieve this purification without it going to the fullest lengths or depths. That's why you start seeing so many turkey vultures every year. The vultures are increasing to try to keep abreast of the increasing dead prey in the environment, as well as to warn us that it could be humans who might no longer be able to survive the environment.

DENNIS: So, as you said before, you are thankful for other people hastening the purification.

MIKE: You have to go with it that way.

DENNIS: Because everything from Nature is a gift, or for a purpose. Even if you don't feel like that yourself, you have to look at it like that.

MIKE: You have to see all aspects of it. Is the work John and I and you are doing delaying the transition?

DENNIS: The overarching principle, I guess, is that things get good, or we can make things good, even if we or some people are making it bad. So, the big one is this, we have to go with the big one, the long-lasting one.

MIKE: Right. And knowing that this has happened in the past helps, although such events have never been identical. The life that remains afterwards, in the next phase, is higher functioning conscious life. Even though difficult at first, they become humble. They also seem to pass along an insight that says we don't want to have to do this again, even though it is probably slated into the structure of Creation. Christians like to say this leads into what they call Armageddon.

DENNIS: So the animals pick up reality in a way more than humans.

JOHN: Oh yes.

MIKE: That's why they don't get affected in the sense of having to face ethical dilemmas. And there's no deceit or hoarding that oppresses other members of their species.

DENNIS: There's no pretense. There's no angle. Well, the animals can be tricky sometimes when they are hunting, I imagine, but not deceitful in the immoral human sense.

MIKE: It's not really deceit. It might be a behavior that the human observes and defines that way, but there is a higher purpose for it.

JOHN: I think when it comes to deer, or comes to almost all the animals, they show that prophecy is coming.

MIKE: That's why the animals speak. . . . You start seeing all the turkey vultures, the crows, the ravens. It is just so clear that the humans are not making big efforts to diminish their toxic inputs. Unfortunately, we all must pay that price. The purification is getting very close. If the Hado'ih [Iroquois False Faces] have left, or are leaving, then it must be time that we prepare for things to really unravel.

Purification is part of the cleaning-up process, and some-times through the cleaning up it really shakes things. It re-ally disturbs a lot of other stuff. Few talk about the part of prophecy in which we will need people who will call back the Kachinas. We need people who will call back the Hado'ih and other protectors.

The Natural Law is higher than anything that you guys can come up with. There will be a future. What are we doing to be ready for that future? . . . And I'm convinced that the people who need to be leaders are not leaders now.

But that's the thing—the people who know what lead-ership is, and should be, can't step out to be leaders now because they'll get destroyed. That system will make them out to be kooks, or worse. But the leaders who will step into that role have to wait for the cosmos and the prophecy to dismantle. I still think there will be some type of event that will provide a better opportunity for enlightened leadership to be launched, and not have to deal with the nonsense from the old system. So whatever that is, or however that comes about, I think the need is to assist the start of the new cycle, to not have to deal with so much of the garbage that is left behind.

How can you be in denial of saying that Nature really knows what it's doing, and what it produces because of what it's doing to benefit every life out there? And the humans ben-efit even though they don't really realize they benefit from it. They could benefit even more, if they started to realize what they are benefiting from.

We're seeing Nature only from the human perspective. Which I don't think is a very accurate depiction of what's really going on. In my opinion, the nonhumans are much higher functioning than humans. You can just watch the birds, how they operate, and that they are constantly watching. Humans don't live life like that—they don't keep an eye out for things, and that's why they end up having experiences that they call accidents to their surrounds or what they are heading into, like climate change, species extinctions, and so forth.

It's sad because Nature has higher functioning life-forms. It's very interesting to observe how the birds are constantly aware. They have to be. I mean your life depends on being constantly aware and humans have lulled themselves into this sense of "Well, that's why we have police." But shouldn't you be aware of your surrounds and potential problems that could arise or situations that could be harmful to you? And yet it doesn't seem to register to most people how this life functions. It's not just the birds—it is all that life out there.

DENNIS: John, you would say the same thing?

JOHN: Yes, definitely.

DENNIS: Birds and animals are higher functioning than humans.

JOHN: Yes.

AWARENESS AND NATURAL LAW

They help the environment, but they also help our souls.
—JONATHAN FRANZEN, "Why Birds Matter,"
National Geographic, April 9, 2019

Because humans, unlike birds, can and do deviate from Natural Law, the environment and society are in trouble. Yet Nature and birds can be the avenue to self and societal correction. Different aspects of the self and of birds can be addressed depending on what is happening in one's own life, and how one thinks about one's own life in relationship to birds, other humans, and Nature. Being open to Nature and the messages of birds means foregoing a self-satisfied, doctrinaire perspective and conceptualizing oneself as directly interactive with the environment. Natural Law needs be the overarching human model for concerns, from minor to major plans and decisions. It provides the practical, philosophical, and spiritual grid for daily life. The model enables one to use Nature to learn about self and others. There is flexibility within, such that no one tells you exactly how or what to think or do, but life is a process of discovery with the Natural Law.

Self-help or spiritual guidebooks sometimes adopt a dogmatic stance, insisting that (for example) the black-capped chickadee is only associated with bravery, and can become a person's totem or messenger only in order to help that person become braver. Such guidebooks tell you what to think and how to conceptualize one's relationship with an animal or bird in a one-dimensional way.

But the paradigm advocated by Mike and John is not dogmatic and incorporates different views. A particular bird's meaning is best uncovered in practice, in full dialogue with the self and the surrounds.

MIKE: We have this model of what we call Natural Law. It is harmonious. Yet people say, "How can you say it is harmonious when they are out there eating each other?" But they are not eating each other in a cannibalistic way.

Do you think the red-tailed hawk hates rabbits? They don't. What is so hard for people to understand is that Natural Law should be something adhered to, and appreciated, and respected, because It is something that supports life.

JOHN: It is working together, that circle of life.

MIKE: The animals and plants don't say, "We have to get rid of this group," or "we have to take care of that." Everything would work well if the humans were not throwing things so far out of balance. And suddenly you see there are years when there are so many rabbits, and there are years when there are so many squirrels.

What happened? We noticed that DDT played a role in that, and the larger birds of prey were affected, and many died off, and so things got out of balance. They try to make their law higher than Natural Law. Humans make laws without reference to Natural Law and life in Nature. Until you begin to see how all the Natural forces work together, and that there is nothing in there that uses politics, religion, or money as a basis for rules or laws. That is the sad part. Humans detach themselves from the model that was a gift to living things out there. Everything that lives out there abides by Natural Law. I struggle trying to comprehend why they cannot see it. It makes me go, "What's wrong with this," when they cannot use this perfect model for basing how we should start to make a system that says this is how life goes on, this is how life should be lived. They wonder why they are having the problems. It is because they never learned what Natural Law was. Natural Law is the greatest gift to all the life on the planet.

There is a story about men and women, and how life was designed. . . . This was presented to the men, how enjoyable life can be, and if life is really that enjoyable, then would you like to live forever? And the men said, "Yes, we would like that." And it was presented to the women, and they said, "No, we would not like to live forever." And it was asked, "Why not?" So all the clan mothers counciled, and they created a big tight circle, and blocked the men's view. And the men heard all this laughing and talking, and they got interested. They asked themselves, "How come the women are all gathered, and huddled, and talking and laughing?" Finally, the men got enough curiosity or courage, and went up to the women. They asked, "What's going on?" And they said that the women had answered the question, "No, we don't want to live forever." And the men said, "Why not? We could live and enjoy each other and have all that fun."

The women said, "OK, we'll show you." And in the circle, they had all babies, little skunks, baby squirrels, all the little animals. The women said, "If we were going to live forever, we will never see new babies, we won't need them. So, we don't want to live forever. It is the cycle."

JOHN: It is a powerful story.

MIKE: It is a delight to hold a little baby, kittens, puppies. The women had them all in the circle. And they had said, "You won't see these anymore; the adults will just live forever." . . . So you see why it is necessary to have those changes, those cycles. It is key. It is important to have that experience, even though at some stage in life, we might think we would rather just live forever. But that is not the reality. It is important to have that new life. New generation.

JOHN: Where did you hear that from?

MIKE: Way back somewhere, they told the story, and it was much longer. It was probably one of the grandmothers telling the story. It made sense.

JOHN: I wonder if it is still being told now.

MIKE: I don't think so.

JOHN: That is sad.

MIKE: Part of it is that it is the stage of life. . . . They don't teach much like that anymore. That's unfortunate. But everybody who experiences it, you know, from the baby birds to the squirrels, other animals when they are orphaned in the spring, you become attached to them. Look at how nice it is to have them close by. That was a good story. . . . When you get into the other stages of life, you say let's see if we can extend this stage as long as you can. . . . There's a lot of good teachings out there. All the old stories, the way they had a purpose, not just entertaining. We are here not just to entertain you. These are some of life's lessons. Pay attention.

JOHN: The stories are more important than the language. That's how I feel. The stories have value.

BEFORE THE TOBACCO OFFERING

DENNIS: Mike, when you make offerings with tobacco, is that with the understanding that there are different parts of the universe that you are invoking, or communicating with?

MIKE: Yes, but it is typically not broken down into segments. Because the wind is in contact with every insect, with every bird, every tree. Every physical thing out there is in contact with the wind. Now, when we put in our request—and it is not really "our" request—I believe we are putting in the request on behalf of all of them. All these living things are included, with whatever you need. We are here to assist, we are here on your behalf, and you offer gratitude, not just for the wind, but for all the living elements. Now you can see [out the window], that this box elder has seeds to distribute. So milkweed, dandelion, pollen, all the elements that need their future generation, the wind is going to carry. Those are the life forces. It says, "Don't worry, the wind is going to take it." It is not up to the humans to make those choices.

I think it is difficult for most people to envision a concept that encompasses everything. Even the song of each bird, whether you hear it or not, is still being carried by that wind. And so it is inclusive, saying, "We're with you, and we want this message on your behalf and on behalf of all those willing to hear it. We want your message to be carried to whoever needs to hear that." So, it works. It really does.

DENNIS: I have another question for you, John. This one is a little vague. Sometimes you hear a bird or something and you get a message. How is that message communicated from the bird to you?

JOHN: It is like the spirit of the bird talking to my spirit, and my spirit is telling me just why this bird came. This is what you need to do, this is how you need to be, whatever the case. That is the only thing I can say.

DENNIS: Is there an invisible thread, is there a force field, is there something, I don't know, something invisible, or perhaps visible, that some people can see, that really transfers through space?

JOHN: You know we talk about this out at the sweat lodge. Just before everyone went in, five or six nuthatches were outside the sweat lodge. They were talking up a storm. One was saying to me, to us, "I have faith in you." That's what I was getting from the birds.

We have the Grandfathers in the middle of the lodge, and we are sitting around, and I say, "Have you ever thought about the umbilical cord that is connected to the Grandfathers, or connected to the Grandmothers? It is almost the same thing with Nature."

Here is something that just happened the other day. I was out getting some of the goldenrod leaves, cutting some of the flowers on top of them. This is going to sound weird. But all of a sudden, the plant said, "I don't like my name." Now I wasn't thinking about the name at all. I was getting the flowers. What I heard was "yellow rod," not goldenrod, which is a name from overseas. And I went, "Wow! All the years I have been picking you all, this is the first time you ever really told me this." I thought it was really kind of sad that it got the name goldenrod because it was supposedly like gold, when it is really yellow. Now where did that communication come from? It is that connection that you just cannot explain.

DENNIS: So, you can't explain it. It is not available to explanation.

JOHN: It is like trying to explain love. How do you explain love?

MIKE: I believe that there are elements that we do not see, that communicate with us all the time, but we are too distracted by whatever is going on right around us. We just don't catch it.

DENNIS: Is there a way to describe the communication that you cannot see, how it actually happens?

MIKE: There are several ways. I am convinced that is not just one place where these connections come from. It is not just your Ancestors. It is also the birds. It is also the trees.

JOHN: People are not paying attention. There are many storms and fires.

MIKE: It's Nature's turn now. I think we are going to feel it. . . . I mean when I see how quickly the weather can turn. It's ramping up.

Near the end of our last discussion, Mike announced that we would need to make a tobacco offering. We needed to let all the birds, all the elements, know that we respected them and to make sure we offended none of the elements of Nature.

MIKE: This is a good opportunity to make a tobacco offering. For any of the other Elements that need to be recognized.

Sometimes during an offering something comes in that needs to be said. If something comes in, I will tell you. Most times you just let it flow. Part of the wind's purpose is to send things, messages to us, and the wind uses tobacco to send messages. Tobacco is unpredictable. But it is always a good thing. Sometimes one has high expectations about communications getting through, but that is the human flaw. But if you word it correctly then the messages are received.

And sometimes it can be disturbing to humans who also need to receive messages. It can take a little time, but the tobacco knows whose behavior needs to be addressed.

JOHN: Well, you know I have smoked with a lot of people. Some Sun Dancers. With Mike you can really feel how real it is. Every time Mike smokes, you do not know what to expect from him. Seriously. Because he is connected. I have seen him just put tobacco in a pipe. Just reaching like this and putting it in the pipe.

MIKE: I have heard some of the Medicine People talk about tobacco and what we use the tobacco for. It is with the offering that we are trying to bridge communication. And if we did not use the tobacco, they said it would take a part of the person who is doing the communication, doing the offering. That is why you designate the tobacco. Tobacco was given that assignment. You are just the one using the tobacco, the fire; it is not really the person who is offering, it is the tobacco. It is recognized that it is not me, it is just that I am using it in the way it was designed to be used.

That too is Natural Law.

For years, I have regularly sought out birds while I sit, walk, watch, and listen outside. But I also often see several kinds right from my large-windowed front porch. My minimal knowledge of birds, my collection of specific bird experiences, and my memory and anticipation of the wide variety of birds here in the warmer seasons provide a foundation that nurtures and enlivens me. Even when I am away from this upstate New York local bounty of bird activity, I bring bird awareness wherever I travel, local or afar, making sure I bring a bird guide. Although I do not use checklists nor belong to any birding groups, this ever-on-the-lookout for birds provides a grounding, a familiar viewscape, literally and existentially.

When I am in cities, traveling, or facing long wintry days, my recall and imagination about birds I know carry me. Especially as spring approaches. I mark the rounds of the seasons, and much of the stability of life, with the birds' comings and goings—daily, monthly, annually.

Birds teach me about myself, about relationship, about purpose. I want to get closer to birds. Some species, like chickadees, and some individual birds, like the phoebe who nests by the side door, like to come close. Others too, cedar waxwings, female cardinals, and crows, sometimes adjust their flights and movements toward me, or in my direction. But more often it is otherwise, as a house sparrow or catbird or bobolink hurries away or behind a tree limb so I cannot gaze at them. I wish there was a way to literally sit down and talk with all the birds. But, I anthropocentrically justify, the birds are going about their own necessary busyness, and business, independent of their proximity to me.

Yes, I depend psychologically and spiritually on the birds, more than they depend on me for much of anything.

I am certain that many birds know my intent to watch them, even from afar, and even when I am about to try to look more closely at them with my binoculars. Or am about to change my position, even when I am making no observable movements. I have experienced this many a time after settling into a stationary, still posture. They often read my investigatory intent, my intrusive stare, and look away.

It has been my experience that once you pay attention to the fact that when you are outside birds know and see that you breathe, stand, walk, sit, or make noise, you realize that they often make decisions and behave with you in mind. A bird's decision to fly close by or distantly, to land, perch, or hop about is carried out with you as a major factor. Moreover, everywhere a person walks s/he is intruding in overlapping bird territories and econiches (often seasonal), most of them established long before modern human occupation. In a human's known or unknown presence near a bird, it is usually no coincidence how close or far the bird continues to fly or position on a branch or on the ground. It also may or may not desire eye contact.

Then I saw a raven flying over. I had previously heard him over the field behind the pole barn. I was the only one around, so he must have intentionally flown overhead right by me. . . .

Then later, in a wide opening in the canopy to the sky, I saw the flight of a lone yellow warbler cross my vision. As I meandered back along the path, I heard wings flapping, like a ruffed grouse taking off. Indeed, it was, and I turned my gaze to the place where the ruffed grouse landed just across the stream in the thicket of willow bushes. For the first time, I got a quick, but excellent direct look at the grouse's face. Many times over the years, I surprised a grouse and then saw their startled, exiting flights, but only from the side or rear of the bird in low flight below the canopy. This was my first eye-to-eye contact with a grouse. It was a very brief interaction, but because we saw each other looking at one another there was an energy exchange, a full human-bird, bird-human recognition.

Another time, another experience etched in my mind's eye was my seeing right into the eyes of a ruby-throated hummingbird as she turned her head to one side, then the other, as she looked through the porch window directly into my eyes.

Perhaps mutually recognized, eye-to-eye contact between bird and human is the unique event that triggers people into deep engagement with birds. Eye contact crystallizes for me the relationship, the importance of another life being and their attractiveness, and takes me deeper into spirit with nonhuman sentient beings. I suppose that my and others' attribution of personalities, meanings, and mind states to specific birds, or birds by species or in general, suggests that they have some human or human-like characteristics. And that humans reciprocally have some bird or bird-like characteristics, such that communication and understanding are exchanged.

Since birds are often the only wild animal life a person can see in city or country, birds can be the most accessible form of nonhuman conscious life. Indeed birds, in much literature and religion, are *the* go-betweens for human and nonhuman, nature and culture, body and soul in transition.

But, of course, birds are very much prey to human activity—hunting, climate change, habitat destruction, tall buildings, wind turbines, cats and dogs. Moreover, and concomitantly, most modern Western human attitudes and philosophies incorporate no *deep* meaning in the appearance and actions of birds. And city dwellers, especially, often lack an implicit or explicit understanding of birds' crucial roles in ecosystem maintenance and the large-scale processes of evolution. We *depend* on birds for ecosystems' survival, development, and harmony with nature.[1]

Birds are crucial, not only in ecological terms for controlling insects and for pollination, for fertilizer, food, and other ecosystem "services," but also they provide visual pleasure as flying beings that diversify and entertain our surrounds.

Birds, flying and alighting in air, sea, and land, stimulate human attention and provide forums of contrast, foreground and background with human movement and awareness. They energize simple and complex human cognitive functions, from recognition of sound patterns to visual discriminations within an almost limitless array of color and shades of color. They also bring shape and melody into our lives. They fill our visual and auditory worlds with unparalleled variety. They surprise and sharpen our sensibilities, ruffle our mental feathers, touch to the quick, and, for some, enchant our souls. Birds incite and excite our imaginations and, bridging earth and sky, provide linkages to higher realms. Intended to divert our attention from routine thought, they are natural church bells, singing bowls, gongs, handclaps. And they can inspirit the ordinary with flashes of insight.

Prior to the many hours I spent with Mike and John in John's house, outside, and in local diners, my ornithology was limited to a generalized theological notion that birds are sacred as part of Creation. I had often only amorphously read into the appearance of birds. I often thank out loud many birds, especially some of my favorite (totem) birds, like belted kingfishers, who for twenty-five years have religiously returned to the creek by my house. I thank them for being around, and for reminding me about them and their importance to my geographic sense of place and to my location as a human in Nature.

But only after extended discussions with Mike and John, tantalizing my thinking about, feeling about, and experiencing birds, did I come to a fuller realization that the sacredness of birds could be even more direct. Meaning that invisible forces some call Spirit, the Creator, God, Nature, the Great Mystery, or other terms or concepts humans use to name the ongoing divine origin and creation of life and birds can operate in specific experiences with a particular bird at a particular time and place. I was opened up, led beyond the traditional literary and aesthetic Western perspective of bird meaning, to explore birds as embodied spirits and mystical presence.

Individuals and groups historically directly dependent on nature for subsistence have often maintained the sanctity of birds and not let human-made goods or advanced technologies divert them from these kinds of knowledge and experience. A bird is a meaningful being in many ways. A bird even can bring the message that there *is* meaning in life, that creation, evolution, and life are to be celebrated and that diversity is the natural order of things. They are biological and spiritual enchanters that infuse life and passion into the biosphere.

My own long-standing immersion among birds in the natural surrounds in the countryside and my association with Mike and John inspired me to research mysticism in ornithology.

Historically, in some European contexts, more commonly in esoteric religious or philosophical groups, as among alchemists and mystery schools, bird traits and stories do rise beyond metaphor, folklore, allegory, or symbol. Birds, even in the West, have been soul-stirring characters and have assumed crucial roles as representatives of higher forces and can act as divine messengers of information from divinity. That is, sacredness in birds occurs independently of a human's attribution of that sacredness to bird events. Birds on their own carry Spirit. It takes intuition to read and experience bird meaning and message, readings and meanings not supervised by a religious official or text.

In ancient Greece and Rome, birds were used in foretelling the future, in divination. Indeed, the English word *ornithomancy*, which is reading omens from the flight, actions, and sounds of birds, comes from the ancient Greek words *ornis* (bird) and *manteia* (divination). Ornithomancy also included reading a message from the gods. It is similar to ancient Roman augury. Augurs were special Roman priests. The word "inauguration" derives from the Latin *inaugurare*, "to take omens from flying birds."

Communion or communication with God via birds is recorded among prophets and Christian saints, most notably Saint Francis, Catholic patron saint of animals. Saint Francis gave a sermon to the birds, and they responded. Yet, such European perspectives generally have not gone beyond the idealization or reification of traits into a true spiritual experience for the layperson. Not as far as Indigenous people's experience, where the sacred and the profane become fully intermeshed in everyday experience, where forces of divinity are illumined in birds.

There must exist a continuum along which individuals and cultures might locate themselves from a thoroughly secular, scientific kind of (Darwinian) ornithology to a most sacrosanct ornithology where humans witness a personalized divinity when they see a blue jay or slate-colored junco.

Historically, a few Western scholars, scientists, or spiritual teachers—usually also mystics, such as Steiner or de Chardin— have considered sanctity to be literally embodied in birds or other nonhumans.[2] This has come to be recognized and embraced by some through experience in Nature and/or study of what some have called "deep ecology"[3] or "spiritual ecology."[4] Yet these terms can remain intellectual Western responses in the sense that the phenomenon has to be named as such, as different from conventional (scientific) "ecology." This is unlike an Indigenous approach, which does not separate ecology from "deep" or "spiritual" ecology. Deep ecology and spiritual ecology are terms used as bases for an engaged or spiritual environmentalism, often borrowing, (sometimes without reference) from Indigenous experience. Sponsel discusses Indigenous peoples as "the original spiritual ecologists."[5]

Some ecopsychologists have attempted to measure the degree to which individuals are sensitive to Nature and have developed tests or scales such as the Spiritual Connection Questionnaire (SCQ)[6] and the Connectedness to Nature Scale (CNS).[7] Such approaches attempt to measure affective states and beliefs, although beliefs and emotions are often conflated. Moreover, people ask and report about beliefs as if they actively choose their beliefs, when the phenomenological evidence is that people only really come to *find* that they have a belief. And cross-cultural comparisons of the notions of belief indicate that "it is very difficult to separate what is properly experience from what is properly belief."[8]

To recognize and use birds as omens, as creatures for human divination, or as bringers or harbingers of meaning and message from a Creator's realm entails a special approach to folk ornithology. One must have been exposed to and be open to certain possibilities. For example, to even the possibility of the existence of a Creator, or to an animism in which ensouled creatures (birds) are able to communicate well beyond their own survival needs. And that personal human consciousness—thoughts, emotions, intuitions—can at least in part be read by nonhuman or noncorporeal entities.

Mystical experience exists. Nonrational, but verifiable-to-the-individual, "clinical," empirical mystical experience. And while humans can teach *about* mysticism, and expose others to such perspectives, one personally must come to know the reality of mystical experience.

My ethno-ornithological method choice here, using extensive verbatim interviews as data, can help get close to direct experience. But you cannot really teach mystical experience. The best ethnographic detail, the best data, comes directly from people. First-hand accounts and reflections help to diminish unnecessary interpretation or mediation, as well as offset the colonial mindset of explanation in which members of one culture analyze members of another culture. It is my hope that the extended interviews and personal reflections in the preceding chapters have provided inklings into the everyday magic and majesty of birds.

NOTES

INTRODUCTION

1. Bastine and Winfield (2011).
2. Boyd (1994).
3. Williams (1976).
4. MacDonald (2020: 15).

CHAPTER ONE. VIEW FROM ABOVE

1. Louv (2008).
2. See also Cajete (2000) and Nelson (2008).
3. Andrews (1996: 207).
4. Flickers also have been important in the practice of the Peyote Religion in what is now the western United States. See Swan (1992).
5. Nelson (1983) also points out how the spring return of migratory birds is a special time.
6. Ojibwe (and Cheyenne and Lakota) believe(d) that "an owl or owl-human supernatural being serves as gatekeeper along the path to the afterlife. . . . This owl being stands at the fork in the Milky Way, the road in the sky that led to the land of the dead, letting some souls pass but condemning others to roam the earth as ghosts forever." Aftandilian (2013: 66).
7. Members of Native clans tend to take on the traits of their clan animals, of which they are usually fond. See Johnston (1976).

8. Why owls have big eyes is one of many traditional Haudenosaunee owl stories. See Beauchamp (1892) and Wilson (1950).

CHAPTER TWO. BIRD MESSENGERS, TOTEMS, LESSONS

1. Grim (2001).
2. Sponsel (2012).
3. Harvey (2006: 3).
4. Harvey (2006: xi).
5. See earlier discussion.

CHAPTER THREE. MESSAGES OF COLOR

1. See Berlin and Kay (1969).
2. Lee (2007: 54).
3. Lee (2007: 53).
4. See Reichard (1990: 20–21) for a discussion of Navaho color symbolism and Navaho sandpainting.
5. Tyler (1979).
6. Hamell (1992: 257).
7. Pritchard (2013: 101).
8. Pritchard (2013: 102).

CHAPTER FOUR. NATURAL LAW AND ORIGINAL INSTRUCTIONS: CONSEQUENCES, CHANGES, CONNECTIONS

1. Keesing et al. (2010).
2. Rosenberg et al. (2019).
3. M. K. Nelson (2008).

AFTERTHOUGHTS

1. See Sekercioglu et al. (2016).
2. See Steiner (1914, 1989); de Chardin (1954, 1956).
3. Naess (1989).
4. Sponsel (2012).
5. Sponsel (2012: 3).
6. Wheeler and Hyland (2008).
7. Mayer and Frantz (2004).
8. Levy-Bruhl (1938: 10).

SUGGESTED READING

Ackerman, Jennifer. 2016. *The Genius of Birds*. New York: Penguin Random House.

Aftandilian, Dave. 2013. "Interpreting Animal Effigies from Pre-contact Native American Sites: Applying an Interdisciplinary Method to Illinois Mississippian Artifacts." In *The Dead Tell Tales—Essays in Honor of Jane C. Buikstra*, ed. Maria C. Lozada and B. O'Donnabhain. Los Angeles: University of California, Cotsen Institute of Archaeology Press, 62–70.

Andrews, Ted. 1996. *Animal-Speak: The Spiritual and Magical Powers of Creatures Great and Small*. St. Paul, MN: Llellewyn Publications.

Ansary, Mir Tamim. 2000. *Eastern Woodland Indians*. Chicago: Heineman Library.

Bastine, Michael, and Mason Winfield. 2011. *Iroquois Supernatural: Talking Animals and Medicine People*. Rochester, VT: Bear & Co.

Beauchamp, W. M. 1892. "Iroquois Notes." *Journal of American Folklore* 5, no. 18 (July–September), 223–229.

Berkes, Fikret. 2012. *Sacred Ecology*, third edition. New York and London: Routledge.

Berlin, Brent, and Paul Kay. 1969. *Basic Color Terms: Their Universality and Evolution*. Chicago: University of Chicago Press.

Berry, Thomas. 2009. *The Sacred Universe: Earth, Spirituality, and Religion in the Twenty-First Century*. New York: Columbia University Press.

Berry, Thomas. 1988 *Dream of the Earth*. San Francisco: Sierra Club Books.

Blanchan, Neltje. 1926. *The Nature Library: Birds*. New York: Doubleday, Doran & Company.

Boyd, Doug. 1994. *Mad Bear: Spirit, Healing and the Sacred in the Life of a Native American Medicine Man*. New York: Simon and Schuster.

Bucko, Raymond A. 1998. *The Lakota Ritual of the Sweat Lodge: History and Contemporary Practice*. Lincoln: University of Nebraska Press.

Bull, John, and Farrand Jr., John. 1977. *The Audubon Field Guide to North American Birds: Eastern Region*. New York: Knopf.

Cajete, Gregory. 2000. *Native Science: Natural Laws of Interdependence*. Santa Fe, NM: Clear Light.

Chapman, Frank M. 1913. *Bird-Life: A Guide to the Study of Our Common Birds*. New York: D. Appleton and Company.

Cherry, Elizabeth. 2019. *For the Birds: Protecting Wildlife through the Naturalist Gaze*. New Brunswick, NJ: Rutgers University Press.

Collins Jr., Henry Hill, and Ned R. Boyajian. 1965. *Familiar Garden Birds of America*. New York: Harper & Row.

de Chardin, Pierre Teilhard. 1959. *The Phenomenon of Man*. New York: Harper.

de Chardin, Pierre Teilhard. 1960. *The Divine Milieu*. New York: Harper.

Deloria Jr., Vine. 1974. *Behind the Trail of Broken Treaties: An Indian Declaration of Independence*. Austin: University of Texas Press.

Emery, Nathan. 2016. *Bird Brain: An Exploration of Avian Intelligence*. Princeton, NJ: Princeton University Press.

Engelbrecht, William. 2003. *Iroquoia: The Development of a Native World*. Syracuse, NY: Syracuse University Press.

Feld, Steven. 1982. *Sound and Sentiment: Birds, Weeping, Poetics, and Song in Kaluli Expression*. Philadelphia: University of Pennsylvania Press.

Fisher, Celia. 2014. *The Magic of Birds*. London: The British Library.

Gaffin, Dennis. 2012. *Running with the Fairies: Towards a Transpersonal Anthropology of Religion*. Newcastle upon Tyne, UK: Cambridge Scholars Publishing.

Greenwood, Susan. 2013. "On Becoming an Owl: Magical Consciousness." In *Religion and the Subtle Body in Asia and the West:*

Between Mind and Body, ed. Geoffrey Samuel and Jay Johnston. London: Routledge, 211–223.

Grim, John, ed. 2001. *Indigenous Traditions and Ecology: The Interbeing of Cosmology and Community*. Cambridge, MA: Harvard University Press/Harvard University Center for the Study of World Religions.

Gross, Lawrence William. 2014. *Anishinaabe Ways of Knowing and Being*. Surrey, UK, and Burlington, VT: Ashgate.

Guss, David M., ed. 1985. *The Language of Birds: Tales, Texts, and Poems of Interspecies Communication*. San Francisco: North Point Press.

Hallowell, A. Irving. 1960. "Ojibwa Ontology, Behavior, and World View." In *Culture in History: Essays in Honor of Paul Radin*, ed. Stanley Diamond. New York: Columbia University Press. (Reprinted in Graham Harvey, ed., 2002: 18–49.)

Hamell, George R. 1992. "The Iroquois and the World's Rim: Speculations on Color, Culture, and Contact." *American Indian Quarterly* 16, no. 4, 451–469.

Harvey, Graham, ed. 2002. *Readings in Indigenous Religion*. London: Continuum.

Johnston, Basil H. 1976. *Ojibway Heritage*. Lincoln: University of Nebraska Press.

Johnston, Basil H. 2001. *The Manitous: The Spiritual World of the Ojibway*. St. Paul: Minnesota Historical Society Press.

Johnston, Basil H. 2003. *Honour Earth Mother*. Lincoln: University of Nebraska Press.

Keesing, Felicia, Lisa K. Belden, Peter Dagzak, et al. 2010. "Impacts of Biodiversity on the Emergence and Transmission of Infectious Diseases." *Nature* 468, 647–652.

Kimmerer, Robin Wall. 2013. *Braiding Sweetgrass: Indigenous Wisdom, Scientific Knowledge, and the Teachings of Plants*. Minneapolis: Milkweed Editions.

Krech III, Shepard. 2009. *Spirits of the Air: Birds and American Indians in the South*. Athens: University of Georgia Press.

Lee, David. 2007. *Nature's Palette: The Science of Plant Color*. Chicago: University of Chicago Press.

Levy-Bruhl, Lucien. 1938. *L'Experience mystique et les symbols chez les primitive.* Paris: Alcan.

Little Bear, Leroy. 2000. Foreword. In Gregory Cajete, *Native Science: Natural Laws of Interdependence.* Santa Fe, NM: Clear Light, ix–xii.

Louv, Richard. 2008. *Last Child in the Woods: Saving Our Children from Nature-Deficit Disorder.* Chapel Hill, NC: Algonquin Books of Chapel Hill.

Lyons, Chief Oren. 2008. "Listening to Natural Law." In *Original Instructions: Indigenous Teachings for a Sustainable Future,* ed. Melissa K. Nelson. Rochester, VT: Bear & Co., 22–26.

MacDonald, Helen. 2020. "The Comfort of Common Creatures." *New York Times Magazine,* May 24, 14–15.

Mayer, F. S., and C. M. Frantz. 2004. "The Connectedness to Nature Scale: A Measure of Individuals' Feeling in Community with Nature." *Journal of Environmental Psychology* 24, no. 4, 503–515.

Mohawk, John. 2008. "A Seneca Greeting—Relationship Requires Us to Be Thankful." In *Original Instructions: Indigenous Teachings for a Sustainable Future,* ed. Melissa K. Nelson. Rochester, VT: Bear & Co., v–vii.

Naess, Arne. 1989. *Ecology, Community, and Lifestyle: Outline of an Ecosophy.* Trans. and ed. D. Rothenberg. Cambridge: Cambridge University Press.

Nelson, Elizabeth Hoffman. 1998. "The Heyoka of the Sioux." In *Fools and Jesters in Literature, Art, and History: A Bio-bibliographical Sourcebook,* first edition. Westport, CT: Greenwood Press, 246–248.

Nelson, Melissa K., and Dan Shilling, eds. 2018. *Traditional Ecological Knowledge: Learning from Indigenous Practices for Environmental Sustainability.* Cambridge: Cambridge University Press.

Nelson, Melissa K., ed. 2008. *Original Instructions: Indigenous Teachings for a Sustainable Future.* Rochester, VT: Bear & Co.

Nelson, Richard. 1983. *Make Prayers to the Raven: A Koyukon View of the Northern Forest.* Chicago: University of Chicago Press.

Nichols Jr., Robert E. 1995. *Birds of Algonquin Legend*. Ann Arbor: University of Michigan Press.

O'Donohue, John. 2004. *Beauty: The Invisible Embrace*. New York: HarperCollins.

Pearson, T. Gilbert, ed. 1936. *Birds of America*. New York: Doubleday & Co.

Pomedli, Michael. 2014. *Living with Animals: Ojibwe Spirit Powers*. Toronto: University of Toronto Press.

Pritchard, Evan T. 2013. *Bird Medicine: The Sacred Power of Bird Shamanism*. Rochester, VT: Bear & Co.

Rea, Amadeo M. 2007. *Wings in the Desert: A Folk Ornithology of the Northern Pimans*. Tucson: University of Arizona Press.

Reichard, G. A. 1990. *Navaho Religion: A Study of Symbolism*. Revised Edition. Princeton, NJ: Princeton University Press.

Ritzenthaler, Robert E., and Pat Ritzenthaler. 1983. *The Woodland Indians of the Western Great Lakes*. Prospect Heights, IL: Waveland Press.

Robbins, Jim. 2018. *The Wonder of Birds: What They Tell Us about Ourselves, the World, and a Better Future*. New York: Spiegel & Grau.

Rosenberg, K., Adrian M. Doktor, Peter J. Blancher, et al. 2019. "Decline of the North American Fauna." *Science* 366, no. 6461, 120–124.

Roszak, Theodore, Mary E. Gomes, and Allen D. Kanner, eds. 1995. *Ecopsychology: Restoring the Earth, Healing the Mind*. San Francisco: Sierra Books.

Rothenberg, David. 2005. *Why Birds Sing: A Journey into the Mystery of Bird Song*. Cambridge, MA: Basic Books.

Roza, Greg. 2003. *The Iroquois of New York*. New York: The Rosen Publishing Group.

Rozzi, Ricardo. 2010. *Multi-Ethnic Bird Guide of the Sub-Antarctic Forests of South America*. Denton, TX: University of North Texas/Universidad de Magallanes, Chile.

Russell, Priscilla N., and George C. West. 2003. *Bird Traditions of the Lime Village Area Dena'ina: Upper Stoney River Ethnoornithology*. Fairbanks: University of Alaska Press.

Sekercioglu, Cagan, Daniel G. Wenny, and Christopher Whelan, eds. 2016. *Why Birds Matter: Avian Ecological Function and Ecosystem Services*. Chicago: University of Chicago Press.

Shenandoah, Joanne, and George M. Douglas. 1988. *Skywoman: Legends of the Iroquois*. Santa Fe, NM: Clear Light Publishers.

Speck, Frank G. 1946. "Bird Nomenclature and Song Interpretation of the Carolina Delaware: An Essay in Ethno-Ornithology." *Journal of the Washington Academy of Sciences* 36, no. 8, 249–258.

Sponsel, Leslie E. 2012. *Spiritual Ecology: A Quiet Revolution*. Santa Barbara, CA: Praeger.

Steiner, Rudolph. 1914. *An Outline of Occult Science*. Trans, and ed. Max Gysi. Rand McNally: New York and Chicago.

Steiner, Rudolph. 1999. *Autobiography, Chapters in the Course of My Life 1861–1917*. Hudson, NY: Anthroposophic Society.

Stokes, John, and Konawahienton. 1996. *Thanksgiving Address: Greetings to the Natural World*. Six Nations Museum and Tracking Project.

Swan, James A. 1992. *Nature as Teacher and Healer: How to Reawaken Your Connection with Nature*. New York: Villard Books.

Tidemann, Sonia, and Andrew Gosler, eds. 2010, *Ethno-Ornithology: Birds, Indigenous Peoples, Culture, and Society*. Oxon, UK, and New York: Earthscan/Taylor and Francis.

Todd, Kim. 2012. *Sparrow*. London: Reaktion Books.

Tyler, Hamilton A. 1979. *Pueblo Birds and Myths*. Norman: University of Oklahoma Press.

U.S. Fish and Wildlife Service. 2011. "Traditional Ecological Knowledge: For Application by Service Scientists." http://www.fws.gov.

Vakoch, Douglas, and Fernando Castrillon. 2014. *Ecopsychology, Phenomenology, and the Environment: The Experience of Nature*. New York: Springer.

Wagamese, Richard. 2016. *Embers: One Ojibway's Meditations*. Madeira Park, British Columbia: Douglas and McIntyre.

Wansbury, Andrea. 2006. *Birds: Divine Messengers*. Forres, UK: Findhorn Press.

Wheeler, P., and M. E. Hyland. 2008. "The Development of a Scale to Measure the Experience of Spiritual Connection and the Correlation between This Experience and Values." *Spirituality Health* 9: 193–217.

Williams, Ted C. 1976. *The Reservation*. Syracuse, NY: Syracuse University Press.

Wilson, Eddie. 1950. "The Owl and the American Indian." *Journal of American Folklore* 63, no. 249 (July–September), 336–344.

Young, Jon. 2012. *What the Robin Knows: How Birds Reveal the Secrets of the Natural World*. Boston and New York: Houghton Mifflin Harcourt.